Polar Bears in the Hot Tub

Their Future
- And Ours -

Arthur Krugler

September, 2018

POLAR BEARS IN THE HOT TUB

Copyright © 2016 by Arthur Krugler

All rights reserved.
Published in the United States

Library of Congress Cataloging-in-Publication Data
Krugler, Arthur
TXu 2-037-818

ISBN-13 978-0692191989 (Krugler Engineering Group, Inc.)

Printed in the United States of America

POLAR BEARS IN THE HOT TUB

Table of Contents

Forwards by Chapters - One through Six	6
Chapter One - North Pole, Home of the Polar Bears	19
- South Pole, Home of the Penguins	31
Chapter Two - Fuel Types and CO2 Concentration	37
Chapter Three - Coal, As a Fuel	60
Chapter Four - Liquid Hydrocarbons	65
Chapter Five - Greenland's Black Ice	71
Chapter Six - The Future - With Understanding and Wisdom	78
Table of References	86

This book, this 'house', was built one board, one shingle, one datapoint at a time. Each piece of data was examined then fitted carefully in place.
It is now my home, it is comfortable, it may endure for a time.

The goal is to ask and answer questions about the greatest challenge of our time:
How do we, the caretakers of the world, predict the future of this world?
Do we really understand our world? Do we understand Global Warming?
Can we control any part of it?
This book presents data, and new views for deniers, alarmists and the 'don't knows'.
   "Lady, you cannot fly to the sun! It is too hot; your rocket will burn up!"
   "Don't be silly. They have it all figured out. We will fly at night"
Many scientists have it all figured out - with a hockey stick and theories and models.

POLAR BEARS IN THE HOT TUB

POLAR BEARS IN THE HOT TUB

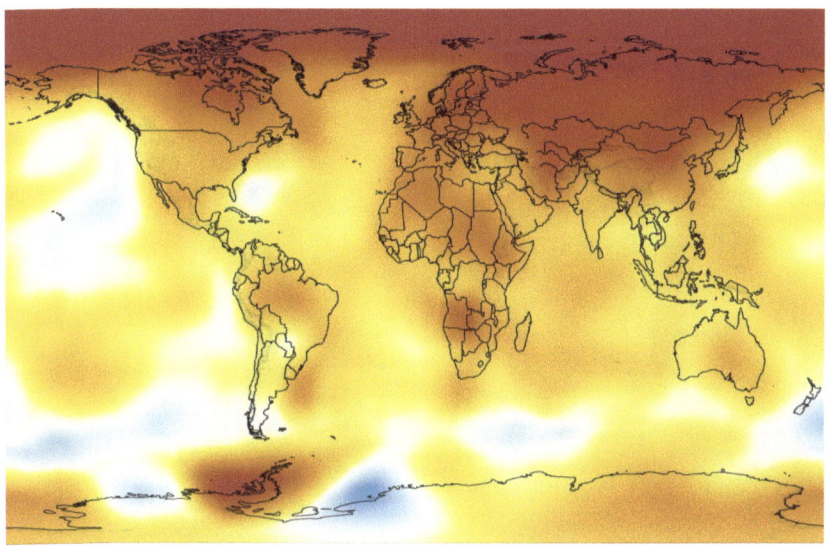

Ref. #1   A world map showing temperature increases in color.

NOAA Data: Air temperature increases from 1960 to 2016.
Red areas, around the North Pole increased 8 to 10 degrees F.
Tan areas increased 1 to 2 degrees and blue areas decreased +/-5 degrees.
Dark red dots north of Norway are volcanic islands increased 15 degrees.
These Arctic Circle islands lie in a straight line 2500 miles long,
 and like the Hawaiian Islands, are also volcanic.
   Why is only the North Pole area so hot when Greenhouse Gasses are uniform?

Richard Feynman has said:
"It does not matter how beautiful your theory is,
  It does not matter how smart you are,
    If it does not agree with experiment, it's wrong"
"The first principle is that you must not fool yourself and you are the easiest person to fool"

   Our earth is one big experimental laboratory - millions of years old.

POLAR BEARS IN THE HOT TUB

FORWARD for CHAPTER ONE - - North Pole; Home of POLAR BEARS

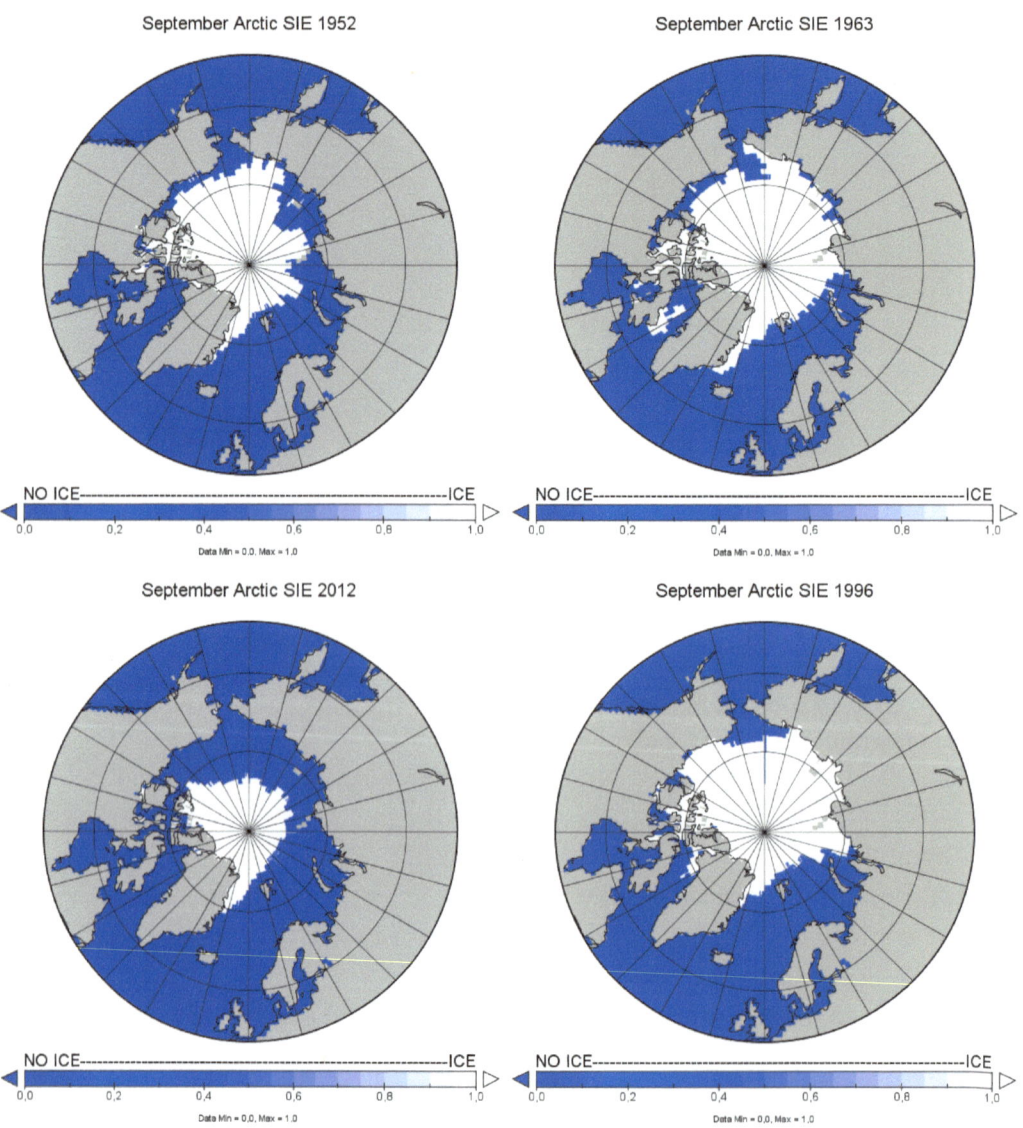

POLAR BEARS IN THE HOT TUB

Observe these NOAA pictures of North Pole Ice Areas.
There are four years:
1952
1963, 11 years later;
1996, 33 years later;
2012, 16 years later.

Note: Area of North Pole Ice (white area) has come and gone since 1952! Polar Bears do not seem to care - population is now "stable" at some 30,000 bears; neither does ice area seem to care. CO_2 had increased steadily after the year 1740 with no consistent reaction from either bears or ice area or ice volume. They often moved in the opposite of predictions.

This next graph of Volume of Arctic Sea Ice is printed vertically. Note the years are from 2003 to 2018 as stated on the graph.

POLAR BEARS IN THE HOT TUB

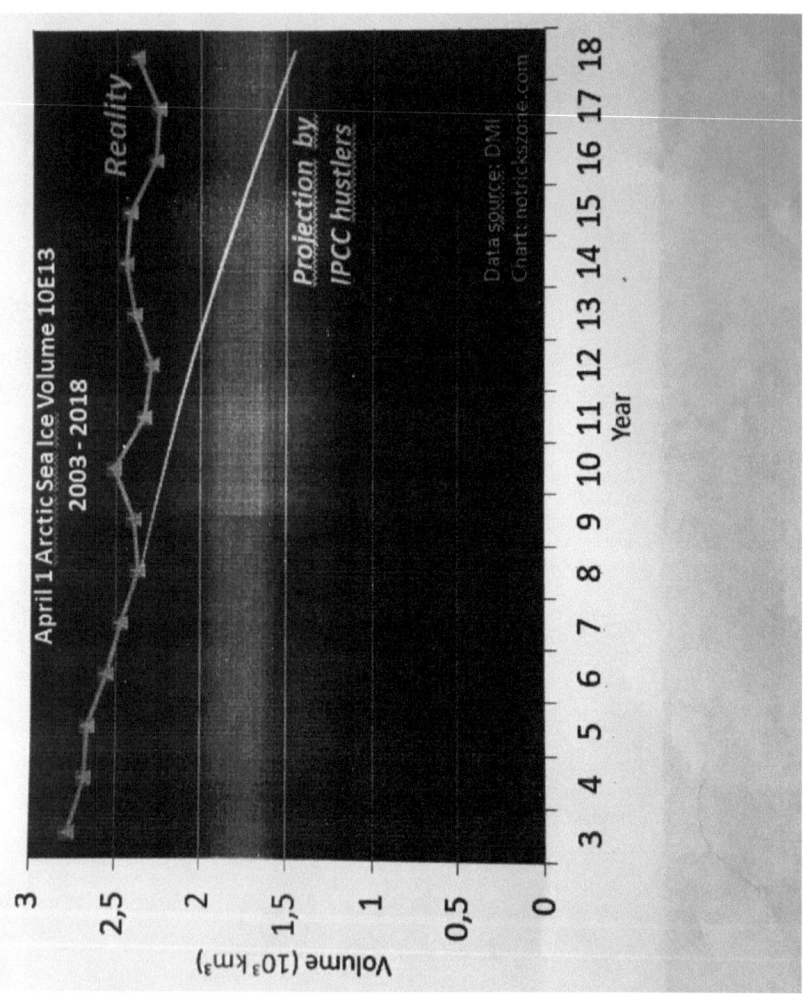

Ref. #3
Ice volume at the North Pole does not respond to CO2 changes: volume comes and goes. CO2 has increased ever faster since 1960 but ice volume since 2008 has not decreased. The Northwest Passage is on hold.

Actually, the ice area and ice volume are reacting to, and are in step with, earthquakes and magma which are now quiet; See the chart in Chapter One showing the rising CO2 concentration, air temperature rising but only after 1980 and no melting of ice until the year 2,000. Neither ice area nor temperature respond directly to CO2 but do respond to quakes and magma.

POLAR BEARS IN THE HOT TUB

FORWARD for CHAPTER ONE, continued - - SOUTH POLE

West Antarctica   << | >>   East Antarctica

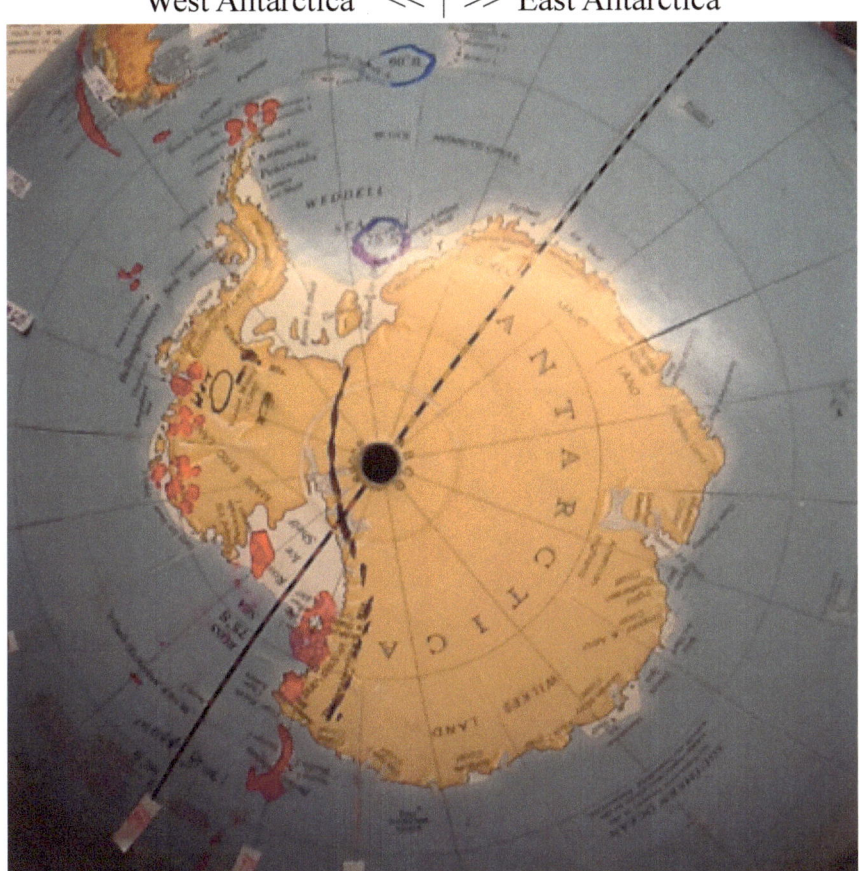

Ref. #4

The red dots locate volcanoes, both active and inactive. They are part of the Ring of Fire which surrounds the Pacific Ocean.

At the South Pole, in contrast to the North Pole, ice does not come and go!

    Ice volume of the larger area of East Antarctica on the right has increased.

    CO_2 has increased ever more rapidly since 1960; but the ice has not responded.

    Sea Ice Shelves and Glaciers of West Antarctica and the Peninsula have not responded either.

    Melting does occur where volcanoes exist beneath two glaciers and where warm water flows beneath ice shelfs.

POLAR BEARS IN THE HOT TUB

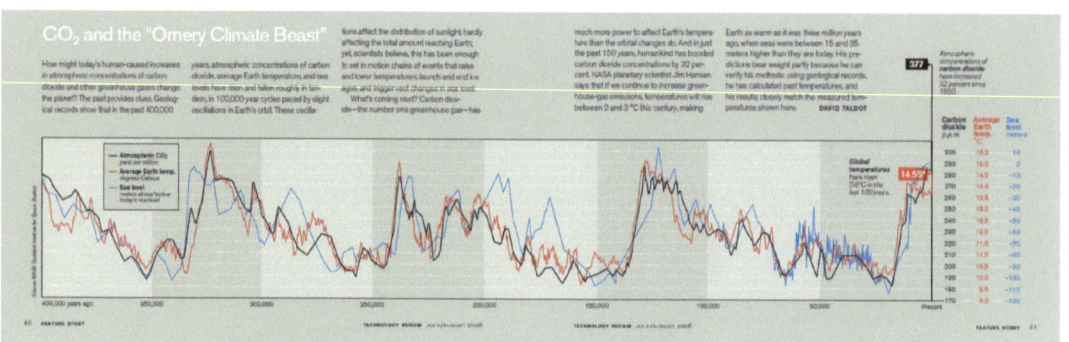

ICE CORE DATA, ANTARCTICA + Mauna Loa

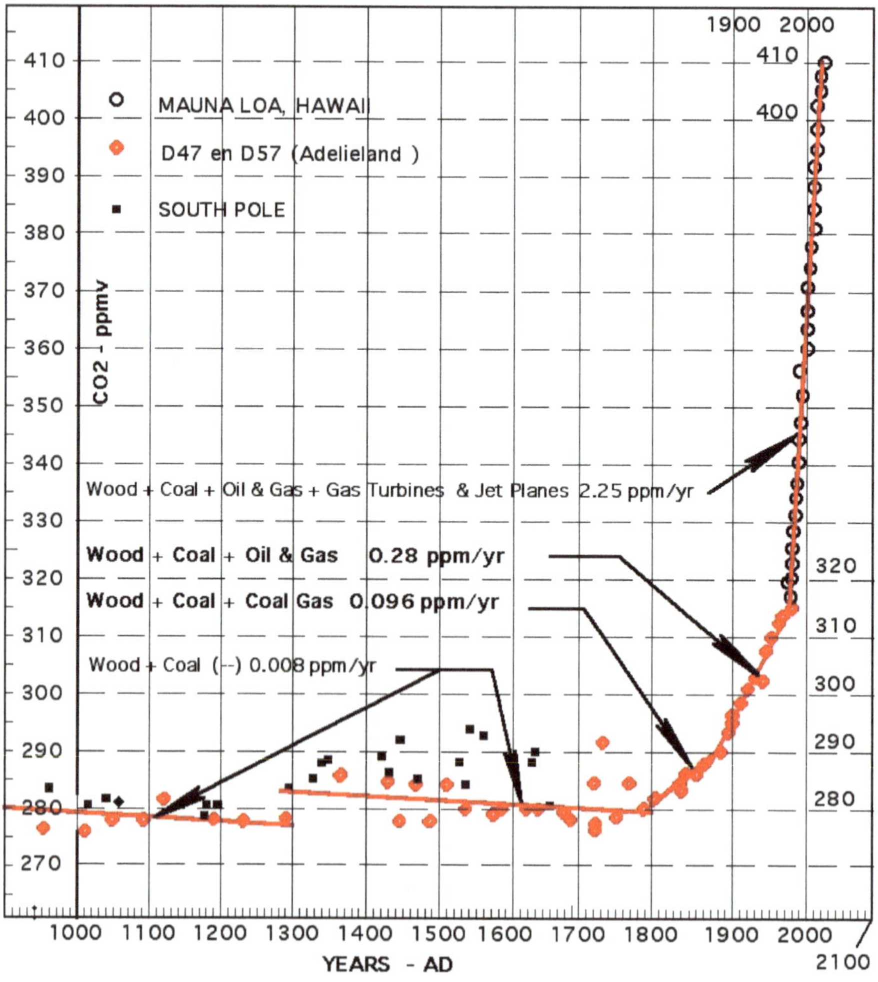

POLAR BEARS IN THE HOT TUB

FORWARD for CHAPTER TWO - THE CO2 HOCKEY STICK

Yogi Berra would advise, "Observe the graphs; there is so much to see".
In the upper graph of 1,000 years, REF. #5, the CO2 hockey stick; note:
1 Before 1760, CO2 was **DECREASING** at 0.01 ppm /per year - YES - **DECREASING**, even though population and the use of coal were **INCREASING**. If unchanged, CO2 concentration today would be 275 ppm.
2 CO2 suddenly started INCREASING after 1760 at 0.1 ppm/year; an abrupt change with a clear reason for the change; production of coal gas.
3 Another abrupt change occurred after 1870, 110 years later, a three fold increase to 0.28 ppm/year, another change with a clear reason; petroleum.
4 Finally an even greater change occurred after 1960 to 2.6 ppm/year, a TENFOLD jump after 90 years of the constant slope of 0.28 ppm/year; turbine engines replacing piston engines for jet airplanes burning jet fuel and power plants burning natural gas. The change from the 1800 slope is 28 times.

The history and chemistry of fuels is discussed in chapter 2 explaining why these changes occurred and when.

In the lower graph, Ref. #6, of 400,000 years of ice core data:
1 The graph shows temperature, CO2 and sea levels for four cycles.
2 Each cycle is 100,000 to 130,000 years long.
3 Usually CO2 concentration follows a temperature rise but not always.
4 Sea levels are erratic but do in a general way follow the 100,000 year cycle.
5 Note that increases in temperature, CO2 and sea level are very steep until a peak is reached in 8,000 years, then the decline to an ice age is quite slow. CO2 concentration did not change the rate of increase or decline.
6 At each peak and each valley, temperature, CO2 and sea level are at nearly identical values. Temperature peaks are declining about 1.4 deg F each 100,000 years. Our world is cooling off.
7 Our Earth has been at a warm peak for 8500 years +/- and is in position to start descending into an ice age except for two reasons;
   - natural nuclear reactors in the earth and
   - carbon nano-particulates from jet and piston engines.

Historical Photograph; Typical of articles on coal plants -Ref. #7

Current Power Plant - Ref. #8

POLAR BEARS IN THE HOT TUB

FORWARD for CHAPTER THREE - COAL AS A FUEL

Coal has been a target of an extensive campaign to have it banned as a fuel. Coal does create environmental problems if burned "as mined" because "coal" is not pure carbon. Sulfur and nitrogen compounds are present in coal and also burn forming oxides which then react with water to form strong sulfuric and nitric acids. These acid gases dissolve in lakes and oceans killing plants, fish and aquatic life. A second reaction, sulfuric acid reacts with salt in ocean water to form bleach which is very effective in killing organisms.

Both acids can be and have been successfully scrubbed from boiler stack gases in the US solving the acid gas problem. Lakes in the US Northeast have recovered surprisingly quickly. Stack gas from coal can be clean.

Burning coal does not produce water vapor and thus the stack gas is heavier than air. This gas tends to settle to ground level for plants and fish to use, rather than rising into the atmosphere to increase CO_2 concentrations.

Replacing power plants that burn coal with plants that burn methane results in increased CO_2 concentration in the atmosphere; this is the very result the "ban the coal" campaign is trying to prevent.

POLAR BEARS IN THE HOT TUB

Ref. #9

Ref. #10

POLAR BEARS IN THE HOT TUB

FORWARD for CHAPTER FOUR - LIQUID FUELS

Liquid fuels are largely a transportation fuel because of high energy density, ease of transportation, storage and low cost. It is used in piston engines for cars as gasoline, diesel engines, and turbines engines both for jet aircraft and some turbine driven power plants which use liquid fuels; most turbines can use either gas or liquid fuel.

Burning liquid hydrocarbon fuels in any engine creates an exhaust gas that is slightly lighter than air; the exhaust gas rises slowly into the atmosphere increasing CO_2 concentrations. Carbon particles are also formed and are carried with the exhaust gas.

While the increased CO_2 concentration from liquid fuels does not appreciably affect temperature, there are two serious problems.

Combustion is not complete and very small carbon particles are produced. Though extremely small, 3 to 60 nanometers in diameter, (0.000,003" to 0.000,060 inches), they are so numerous they affect both temperature and the melting of snow and ice by absorbing sunlight. Note, few jet planes create as much smoke as shown in the photos but they all create particulates.
While in the air the particles heated by the sun, heat the air around them.
Once settled on ice, the particles absorb sunlight and melt the ice.
Once settled on land, both the land surface and the air are heated.
Researchers are now reporting on the impact of carbon particulates on global warming. They have a significant impact while CO_2 has near zero.
Reducing particulates should be major and immediate undertaking.
One approach is to use fuel made from coal gasification; the FT process.

The second problem is sulfur and nitrogen compounds. Sulfur burns much like carbon but produces a very strong acidic gas and Sulfuric Acid.
Nitrogen also reacts with oxygen to produce oxides and Nitric Acid
Both are absorbed in water lowering pH and killing aquatic life.

POLAR BEARS IN THE HOT TUB

In contrast, CO2 is also absorbed by water but there is an equilibrium; at any temperature and pressure there is a maximum CO2 concentration.
CO2 is necessary for all life; slightly higher concentrations are useful.
Reducing sulfur to a very low minimum should be a global priority.
Nitrogen should be carefully studied as it is a useful fertilizer and may not kill aquatic life.

Photo of miles of Greenland's "black ice".

Ref. #11

POLAR BEARS IN THE HOT TUB

FORWARD for CHAPTER FIVE - GREENLAND'S BLACK ICE

Climate alarmists point to the large amount of ice on Greenland which is melting at increasing rates and flowing to the oceans. Rising sea levels would cause major problems for coastal seas and low lying islands.

Greenlands black Ice is caused by soot and the soot must arrive airborne; scientists agree.
Do you suppose this soot settles only on Greenland?
Or would airborne soot also settle everywhere on earth? Land and Sea?
Does it affect melting of ice? Yes, black soot absorbs warming sunlight.
Does it affect global air temperature? Yes indeed.

Black soot on Greenlands ice is reported to be "soot, algae, bacteria and other micro life" but a chemical analysis has not been presented on the web.
This is amazing - vital information is not available.

Black soot likely has only two sources; both from the burning of liquid and gas hydrocarbons. Chapter 5 contains an article on fuel vs particulates.
Requiring these fuels to contain oxygen compounds instead of current diesel and jet fuel would reduce black ice and reduce melting of Greenlands and other ice.
Yes, soot does affect both melting of ice and air temperature.

POLAR BEARS IN THE HOT TUB

FORWARD for CHAPTER SIX - THE FUTURE - WITH WISDOM

CAN THE FUTURE OF POLAR BEARS BE PREDICTED?
CAN THE FUTURE OF ICE AND CLIMATE BE PREDICTED?

What about predicting our future ? The future of the planet?
Can we, the people, do anything about that future?

Chapter six is a discussion of the data, and resulting changes.
With understanding comes the ability to predict trends and outcomes.
As Yogi Berra said, "It is difficult to make predictions, especially about the future". But he also said, " You can see a lot by observing".
There is sufficient data for observations that lead to understanding, and with that the ability to make predictions, after which, wise decisions can be made.

Action on energy efficiency and green energy can reduce the amount of fossil fuels burned. This is desirable whenever economical.

There is an economic limit on the percentage of intermittent power if dependable always-available power is required in tomorrows world. Providing that reliable power with total solar would require 4 full size solar plants, three full size battery storage plants and a full size operating gas plant; seven 'solar' plants plus the single gas plant. Very high costs are inevitable.

Actions being taken to reduce CO_2 will have minor effect on temperature or ice melt. Changing diesel/jet fuels can immediately affect both temperature and melting ice by reducing particulates. The engines also need study.
Finally, our world is entering an ice age with serious impacts on the future.

This dissertation is data and observation driven; not based on opinions.
Nor is it based on computer programs, forcing functions, radiation, albedo, earth movements or other cycles; they are misleading.

Polar Bears in the Hot Tub

CHAPTER ONE - THE NORTH POLE

POLAR BEARS are COOL
They live on top of the world.
They have it all to themselves.
Northern Lights provide wide screen entertainment.
No people, No cars.
Few other animals; None as powerful.
No trees, No bugs, No noisy crows.
For dinner, Seals come to them while they wait.
All summer, large swimming pools, everywhere.

BUT, ARE ALL THOSE SWIMMING POOLS BECOMING HOT TUBS??

Polar Bears – HISTORICAL DATA
Life was especially great until 1900 when they numbered about 25,000 bears.
Then long range rifles with telescope sights were invented.
Bear population dropped significantly to between 5,000 and 10,000 bears.
Then, between 1956 and 1994 Northern nations prohibited hunting.
From 2008 to 2017 the population of Polar Bears rebounded and today there is a stable population of about 31,000 bears. They live about 15 years so 2,000 must die each year.
Polar Bears – Facts & Myths; Susan J Crockford - Ref. #12

POLAR BEARS – ICE & PUBLIC CONCERNS

After 1990, people, concerned people, began showing up in the Arctic, but armed with cameras, airplanes and satellites in the sky. Satellite photos showed that ice at the North Pole was melting! Pictures were published; concerned people started drawing conclusions!

Many research papers were published on causes!

"Ninety-seven percent" of selected scientists agreed; "greenhouse gasses are the cause"!

This is a man-made Catastrophe!

Polar bears will die from lack of ice and food!

The planet will die from hot weather; we will all die!

Burning fossil fuels must be stopped to eliminate CO2!

International Conferences were held in Japan and France! Leaders promised to save the polar bears and us!

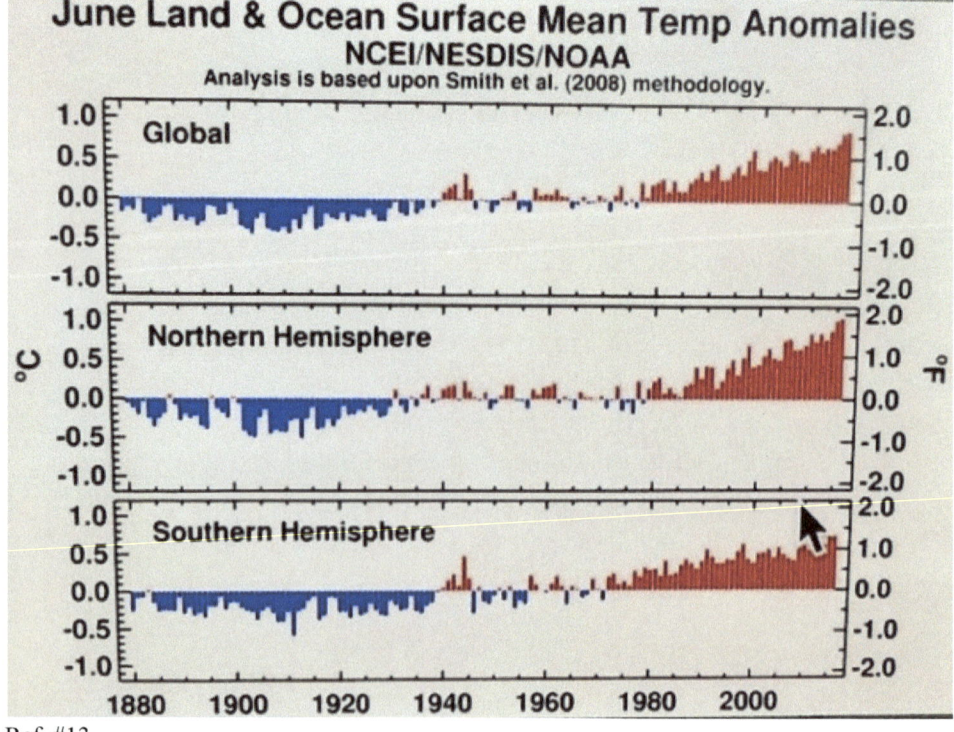

Ref. #13

POLAR BEARS IN THE HOT TUB

The chart is of temperature changes from 1880 to 2016. Deg F scale on right.

Note: There was no increase from 1880 until 1980 while CO2 was increasing. Why did CO2 not cause temperature to rise for 100 years while CO2 was rising?   One hundred years of coal, oil, natural gas and two wars.
In fact between 1880 and 1940 there was a cold spell.
Then temperature started to rise suddenly in 1980. Why then?
CO2 levels were rising continually after 1740.
What if it's not about CO2? Or about us?
What if it's more than just the burning of more fossil fuels?

POLAR BEARS – EARTHQUAKES BENEATH THEIR FEET

Who is not aware of the volcanic eruption of Mount Saint Helens in 1980? The same year the graph shows temperatures began to rise.
Volcanologists and geologists 'listened' to the earthquakes and predicted it would erupt and be fatal for anyone near the mountain.
They established a "RED ZONE" around the mountain and ordered everyone within it to evacuate.
They established an outer "BLUE ZONE" where only permitted people could go. It was presumed "safe".
Three people died in the red zone.
Three died in the "blue zone" with permits to be there.
Fifty-one other people died farther away, outside the blue zone which was considered 'safe'.
Many lives were saved because scientists listened to the earthquake warnings of rising magma.

Earthquakes under the Arctic Ocean have been active and talking to us after 1975 but became almost inactive after 2016.
They are telling us about white hot magma under the Arctic Ocean and it's volcanic Islands. Should we not listen to understand melting Arctic Ice??

POLAR BEARS IN THE HOT TUB

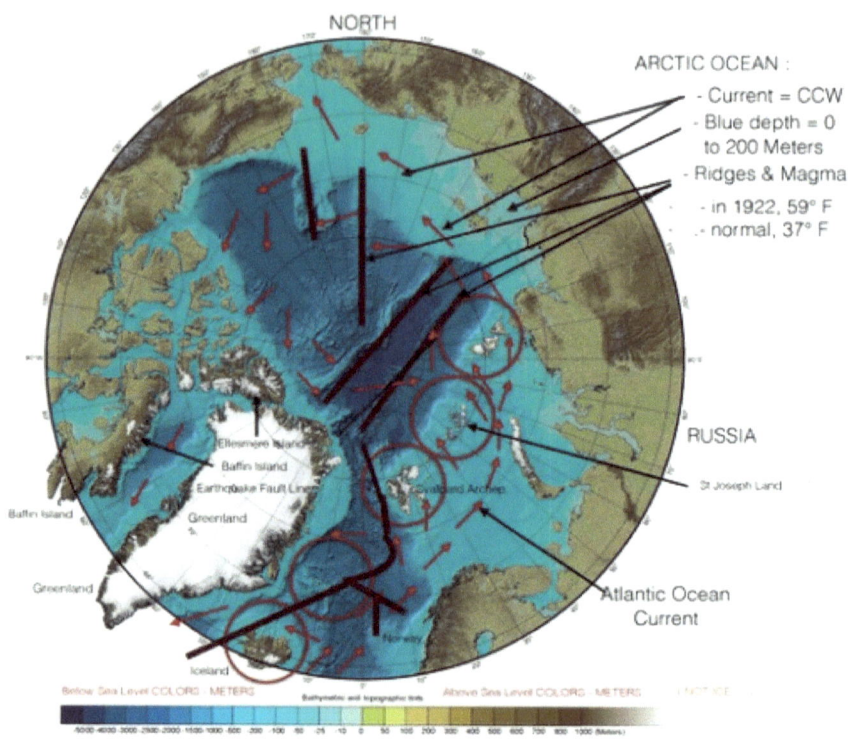

Ref. #14

Earthquake records can be found at earthquaketrack.com where time, location, magnitude and depth are available for many selectable regions. Earthquakes accurately identify areas of rising hot magma.
Like the Hawaiian Island Chain, the volcanic islands of Iceland, Jan Mayen, Svalbard and Russia's Severnaya form a straight line chain of active volcanic islands over 2500 miles long. There is also a parallel undersea ridge some 2600 miles long with similar volcanic activity.

Further, there are active branches to the undersea ridge.
Total length is over 7,000 miles of active earthquakes and rising magma. They are shown as red lines in the picture of the North Pole.

POLAR BEARS IN THE HOT TUB

If the width of this active ridge were known the area could be calculated. One report estimated the width as up to 150 miles, much wider than earlier estimates.
If we add the vertical area of the cracks and channels, the area is truly enormous. That might mean over 1,000,000 square miles of hot seafloor to heat THE POLAR BEAR TUBS.

Since 1980 numerous earthquakes, indicators of volcanoes and rising magma, have been recorded under these Islands and under the Arctic Ocean.

The graph on page 24 has been created to show the lack of response by melting ice to rising CO_2 levels.
It also shows the response of melting ice to earthquake and magma activity. Please note the triangles and dots at the bottom of the graph; each denotes a measure of earthquake energy, but not the amount of heat available from magma to melt ice. These quakes are caused by a spreading seabed, not plate movements. The triangles between 1960 and 1980 indicate near zero quake energy.
Dots between 1980 and 2016 indicate many quakes, each dot indicates the total earthquake energy for that year from many earthquakes.

Earthquake records show almost no earthquakes in the Arctic region for the periods before 1980 and after 2016.
The quakes started in 1980 and then diminished to 1 in 2016 and 2017. Magma is not rising now and the rocky seabed is cooling! If earthquakes do not return, ice will very likely return. Realize that it will likely take several years for the magma and the rock heated by magma to cool and the ice to fully return.

WILL THE HOT TUBS GET HOTTER? OR FREEZE?
The chart on the next page, Ref. #15 is very helpful in understanding the data. It was created to visualize the connections between CO_2 concentration and air temperature, the area of winter ice, area of summer ice, and earthquake activity after 1960.

POLAR BEARS IN THE HOT TUB

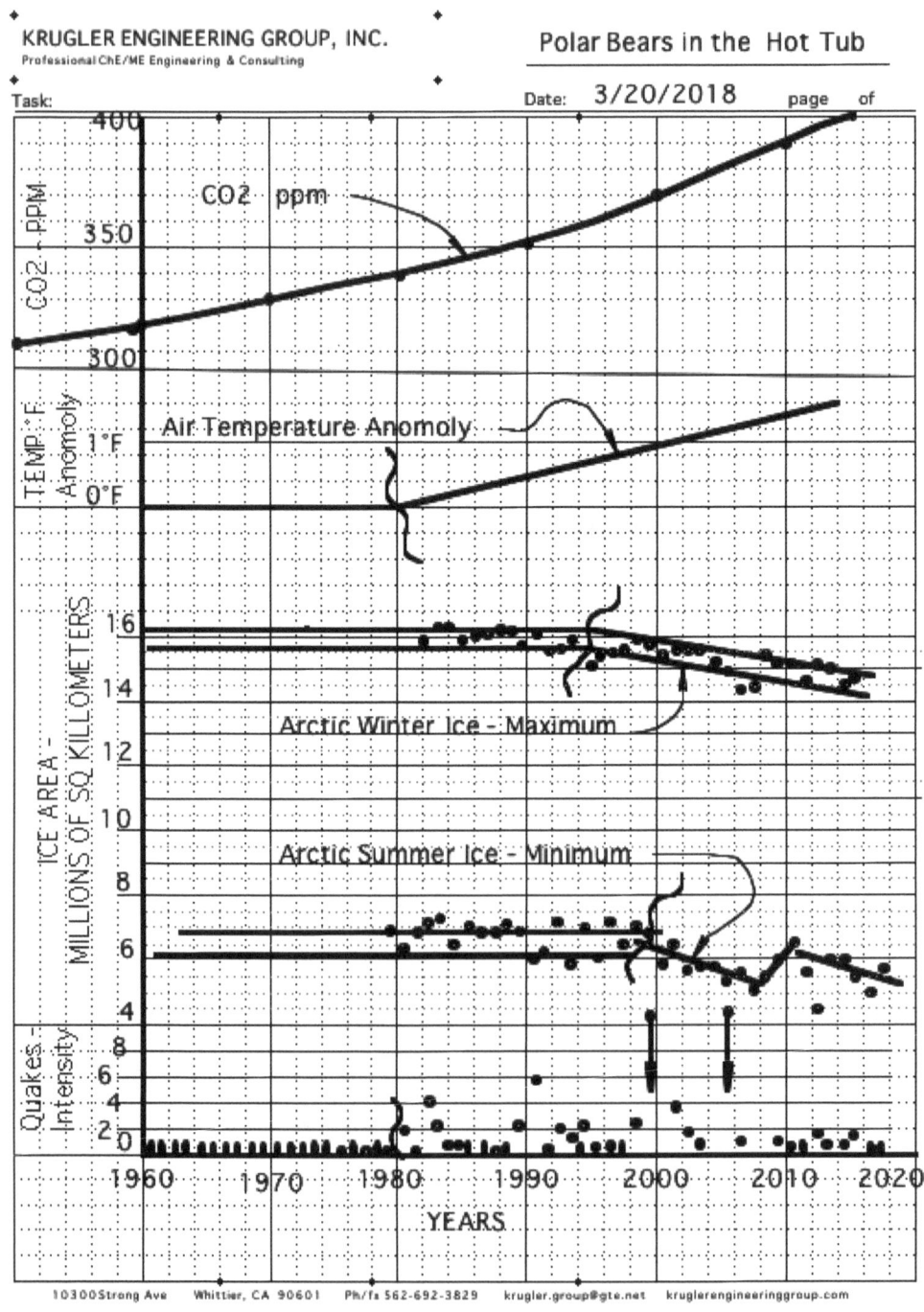

POLAR BEARS IN THE HOT TUB

1 The top line shows CO2 continuing its steady rise.
The next line, temperature, did not rise until 1980!

2 Ice areas did not start to decrease until 1985 after earthquakes indicated magma activity.

3 Temperature did not rise until 1980. Actually, in 1880, 100 years earlier, the air temperature was the same as in 1980 as shown on the NOAA chart, page 23. Clearly, air temperature and ice are not responding to CO2 but a different energy source, which is the rising magma!!!

4 Winter ice area decreased only after 1990, 10 years after temperature increased!

5 Summer ice area then started decreasing in 1998, 18 years after rising temperatures! Summer ice did not respond to CO2 at all for 18 years of temperature increases!!

6 Thinner winter ice, 1 to 3 feet thick, responded more quickly to magma and earthquakes.

7 Very significantly, ice melted along the Russian shore but not along the Canadian shore!! See item 12 below.

8 Thicker summer ice area, up to 10 feet thick, responded 16 years later.-

9 Norwegian expeditions in 2005 and 2008 at 73 deg. North latitude on the Mid-Atlantic Ridge near Loki's Castle north of Iceland, reported numerous underwater volcanoes and thermal steam vents measuring up to 570 degrees F. The magma is there!!

10 [Another Norwegian expedition in 1922](#) studied unusual melting of sea ice and glaciers north of Svalbard 100 years ago. See a following pages for a copy of the published report. Note; accurate records of global earthquakes were not available 100 years ago.

11 In 2013, [Dr Jorge of Norway reported the discovery](#) of an underwater volcanic ridge between Iceland and Svalbard, 1,500 KM long equal to the distance from San Francisco to New York City.

http://www.acadiau.ca/~raeside/quizzes/quiz-11clue.html
ttps://wattsupwiththat.com/2013/08/02/hot-times-near-svalbard-volcanic-range-discovered/

POLAR BEARS IN THE HOT TUB

The 40,000-Mile Volcano By WILLIAM J. BROAD Jan 12, 2016.

A major project in the Pacific Ocean off the West Coast is monitoring a steaming ridge full of living and mineral wonders on the ocean floor circling Earth.   Ref. #16

The New York Times

Note the section on underwater volcanoes and thermal vents.
There can be no doubt that there is a long and large area of magma similar to other areas. See hot magma beneath the island chain.

Satellite photos, page 7, show the ice is melting along the Russian shores but not along the Canadian shore. This is just as it should be! See photo below.
Seawater of the Arctic Ocean is circulating in a counterclockwise direction.

POLAR BEARS IN THE HOT TUB

This means it flows over the earthquake zones and gets hotter, then flows along the Russian shore melting ice and cooling off. By the time this heated seawater flows under the ice north of Canada it has cooled off and can no longer melt ice.
There can be no doubt that there is a major source of heat under the Arctic Ocean north of Svalbard and Russia!
Earthquakes are telling us the Magma had been rising under the Arctic Ocean after 1980. And, as for the future, these earthquakes have largely stopped – the magma is not rising now. If the magma is not rising; this source of heat will disappear.
No help needed from much less powerful human beings!

There is a growing interest in earthquake-magma-volcano fields by earth scientists.
Dr Tolstoy commented that magma may be affecting earths temperature.
Chapter two will explore the history of CO2 concentration in our atmosphere.
The rise is not as simple as "we are burning more so it is rising."
Nor will CO2 stop rising if we stop burning coal and convert to wind/solar.

Ref. #17

POLAR BEARS IN THE HOT TUB

Copy of Arctic Temperature Report; Confirmed by 'Snopes'

IT HAPPENED 100 YEARS AGO AND IS HAPPENING NOW!
MONTHLY WEATHER REVIEW
November, 1922
By George Nicholas Ifft

The Arctic seems to be warming up. Reports from fisherman, seal hunters, and explorers who sail the seas
about Spitzbergen and the eastern Arctic, all point to a radical change in climatic conditions, and hitherto
unheard-of high temperatures in that part of the earth's surface.
In August, 1922, the Norwegian Department of Commerce sent an expedition to Spitzbergen and Bear Island
under the leadership of Dr. A. Hoel, lecturer on geology at the University of Christiania. Its purpose was to
survey and chart the lands adjacent to the Norwegian mines on those islands, take soundings of the adjacent
waters, and make other oceanographic investigations.
Ice conditions were exceptional. In fact, so little ice has never before been noted. The expedition all but
established a record, sailing as far north as 81° 29' in ice-free water. This is the farthest north ever reached
with modern oceanographic apparatus.
The character of the waters of the great polar basin has heretofore been practically unknown. Dr. Hoel
reports that he made a section of the Gulf Stream at 81° north latitude and took soundings to a depth of 3,100
meters. These show the Gulf Stream very warm, and it could be traced as a surface current till beyond the
81st parallel. The warmth of the waters makes it probable that the favorable ice conditions will continue for
some time.
In connection with Dr. Hoel's report, it is of interest to note the unusually warm summer in Arctic Norway
and the observations of Capt. Martin Ingebrigsten, who has sailed the eastern Arctic for 54 years past. He

says that he first noted warmer conditions in 1918, that since that time it has steadily
gotten warmer, and that to-day the Arctic of that region is not recognizable as the same region of 1868 to
1917.
Many old landmarks are so changed as to be unrecognizable. Where formerly great masses of ice were
found, there are now often moraines, accumulations of earth and stones. At many points where
glaciers formerly extended far into the sea they have entirely disappeared.
The change in temperature, says Captain Ingebrigtsen, has also brought about great change in the
flora and fauna of the Arctic. This summer he sought for white fish in Spitzbergen waters. Formerly great
shoals of them were found there. This year he saw none, although he visited all the old fishing grounds
There were few seal in Spitzbergen waters this year, the catch being far under the average. This,
however did not surprise the captain. He pointed out that formerly the waters about Spitzbergen held an
even summer temperature of about 3° Celsius; this year recorded temperatures up to 15°, and last
winter the ocean did not freeze over even on the north coast of Spitzbergen
With the disappearance of white fish and seal has come other life in these waters. This year herring in
great shoals were found along the west coast of Spitzbergen, all the way from the fry to the veritable great
herring. Shoals of smelt were also met with.

End of Norwegian Report
Ref. #18

POLAR BEARS IN THE HOT TUB

POLAR BEARS IN THE HOT TUB

SOUTH POLE - ICE but NO POLAR BEARS

Environmental Scientists are very concerned about the immense amount of ice contained on the very large Antarctica continent. Picture a block of ice two miles thick and one and 1/2 times as large as the United States.
Should all the ice melt and flow into the oceans, scientists calculate sea levels would rise 216 feet. Since most of humanity lives near ocean shores, mass migration to higher land would be required and it would be catastrophic.
Major action is already being taken in the United States to prepare for rising sea levels.
    Ocean front cities are preparing for higher shore erosion. Ocean front planning commissions are requiring new projects to raise the site by adding fill dirt, as much as six feet, before building new construction.
    The US Navy is planning for new harbors as existing Navy yards would be flooded.
    States bordering the ocean are projecting sea rises and spending money planning for major projects and major expenses.
    States, California included, are mandating sustainable cities and 'green' energy to halt global warming.
    These are life changing mandates very disruptive and expensive.

BUT:
What if ice is actually returning to the Arctic and Antarctic?
    There is daily data at: " Charctic Interactive Sea Ice Graph | Arctic Sea Ice News and Analysis. " Paste this link into your browser and select the chart.

    Take time to select the recent years 2012, 2015, 2016 and 2017 with 2012.
    Note: each year has more summer ice than 2012. How can that be if CO_2 which is increasing so rapidly and is predicted by climate models to cause ice to melt??
What if ice is actually increasing as reported, on large East Antarctica to offset any loss in the West?
What if we have actually already entered the next ice age?
What would weather be like during a descent into an ice age?
 - Would we not see more winter snow, and heavy summer rain?
 - Would we stop calling it "severe weather caused by climate warming"?
 What if all this effort and expense is unnecessary, wasteful and actually harmful?
What if all the effort to reduce atmospheric CO_2 is pure waste of time and money?
Our search for more answers is not about whether temperatures have increased since 1980. Temperature has probably increased as shown in the NOAA graph.

POLAR BEARS IN THE HOT TUB

Nor is our search for reasons about the melting of Arctic Ice between the years of 2000 and 2012. The measurements of temperature and ice area may be assumed to be correct for that short period of melting ice.
Nor is there doubt about the very rapid increases in CO2 concentration since 1990.

What then is this chapter about? Actually?
Why has the temperature, as reported by NOAA, increased by 10 to 15 deg. F along the Antarctic Peninsula but other areas rose only about 1.5 degrees.
Why has the ice melted on only two mountain glaciers on West Antarctic?
Why are there about 127 known volcanoes along West Antarcticas' coast where ice is melting but none in all of vast East Antarctica where ice is increasing?
Why are there visible cracks in the West Antarctica ice shelves that are melting?
   If ice is melting from the surface due to warmer air, engineering calculations predict cracks on the bottom of the ice which would not be visible from above, but if melting occurs from warm water beneath the ice, the cracks would occur and be visible from above; as they are.

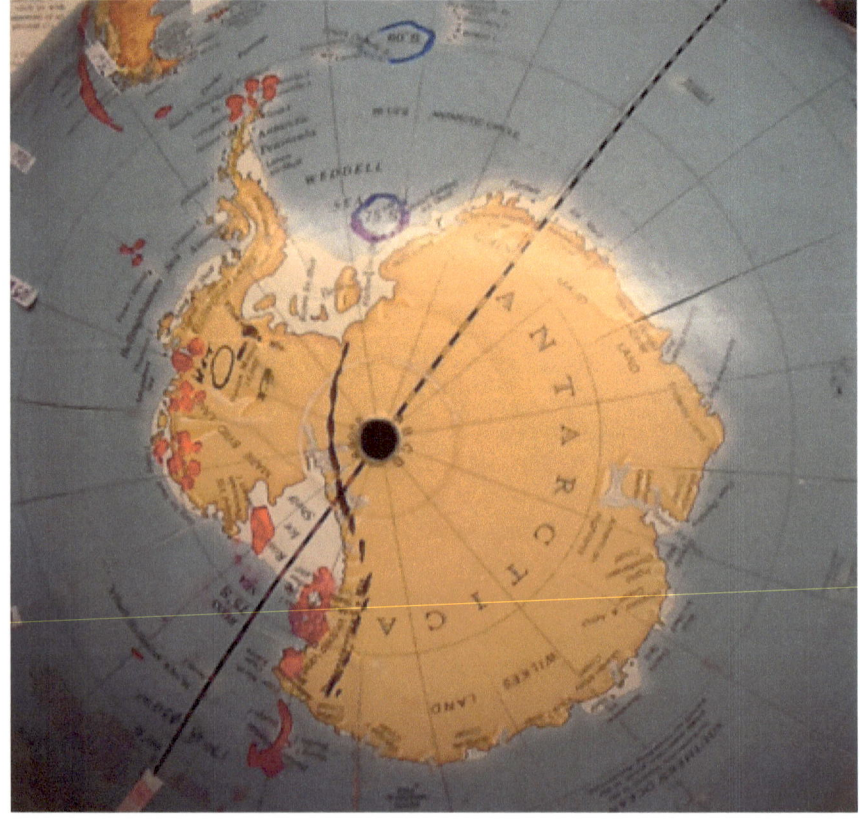

POLAR BEARS IN THE HOT TUB

Lets begin with Ref. #19, a map of Antarctica. Then check out a list of volcanoes.

Notes for the Map of Antarctica:

The longer black wavy line near the South Pole follows a mountain range that divides East and West Antarctica.
East Antarctica (the larger area) is to the right, West Antarctica is to your left.
Antarctica's Peninsula extends up from West Antarctica toward Chile.
Two bays contain major ice shelfs; the Ross Ice Shelf and the Ronne Ice Shelf.
The red dots locate 36 volcanoes which are listed in a table at the end of the Chapter .
Mount Erebus on West Antarctica is actively emitting gas and steam; others are not erupting.
Four other volcanoes on West Antarctica, erupted in 1876, 1892, 1905 and 1970.
East Antarctica, in contrast, has a single dormant volcano located 300 miles offshore.

THREE AREAS
EAST ANTARCTICA - WEST ANTARCTICA - ANTARCTIC PENINSULA

ANTARCTICA INFORMATION
Penguins, over 10 million of them, choose Antarctica as their home.Researchers are concerned about the future of these excellent swimmers.There are 70 research stations studying and reporting on penguins, climate and ice.
The concern originated with rising CO_2 levels and rising temperatures.Antarctica has totally baffling questions if CO_2 is assumed to cause both rising temperature and melting ice.
But first, some facts about Antarctica, the "Down Under Continent".
　　While North Pole regions have thousands of polar bears none exist on Antarctica. Mammals, in fact, do not live on the continent.

　　At 5.4 million sq. miles it is almost twice as large as the United States' with 3 million sq. miles.

POLAR BEARS IN THE HOT TUB

East Antarctica is much larger (72% of Antarctica's' area) and reported to be gaining ice.

West Antarctica, 18% of the area, has numerous volcanoes and is the main area of alarm.

Antarctica is a continent resting on rock, while most of the North Poles Arctic Ocean Ice floats on sea water.

Under the snow lies rocky mountains and a mountain range dividing the continent.

Snow and ice depth over the continent ranges up to 2 miles thick.

Calculations have been made to indicate seas would rise 216 feet if all the ice melted.

Shelfs of West Antarcticas' sea ice are reported to be melting.

Ice cores reveal over 1.5 million years of history (CO_2, sea level and air temperature).

National Geographic's Dec, 2015 map of air temperature changes, show air temperature DECREASES of up to 5 degrees over some sea areas surrounding Antarctica.

This map also shows a 3 degree F increase offshore of West Antarctica, while the air temperature rise over snow covered eastern area is about 1.5 degrees F, similar to the entire planet.

THE RING OF FIRE

1 The Pacific Ocean is surrounded by a "Ring Of Fire".
2 Location of the Ring is well established:
 – 90% of the worlds earthquakes occur on the Ring.
 – 75 % of the worlds active volcanoes are located on the Ring of Fire.
 – Three of the world's largest recent eruptions occurred on the Ring of Fire.
3 The Ring extends south along Chiles' west coast to West Antarctica's Peninsula and then farther southward offshore of West Antarctica before turning northward toward New Zealand.

This chain of volcanoes and magma along West Antarctica's shore is 3600 miles long but width is not known; a width of 100 miles would provide 360,000 square miles of hot seabed.

For comparison, Los Angeles to New York is only 2500 miles;
4 There is one active and 35 inactive volcanoes on the peninsula and shore of West Antarctica.

One inactive volcano is located 400 miles off the coast of East Antarctica. (66°48'S,89°11'E)
5 The volcanoes were created by hot magma which still exists beneath them.

POLAR BEARS IN THE HOT TUB

DISCUSSION

1. West Antarctica is shown by the National Geographic map to now have 3 degrees higher air temperature than 54 years ago; the rest of Antarctica has risen 1 degree or less. Ocean areas around Antarctica are shown to have cooled as much as 5° degrees Fahrenheit.
2. Ice shelfs and glacier are melting in the Western Antarctica region.
3. Volcanoes are shown on a map of Antarctica by dark dots. See page 14.
   Note that most lie off the west shore of West Antarctica under sea ice.
   A table is attached listing all the volcanoes located in Antarctica.
4. Sea rise would be minimal from melting ice; ice shelfs are floating, Coastal ice is minimal.
5. While there is not sufficient data to calculate the amount of ice the magma might melt, science cannot be "settled" until data is available and able to predict melting ice shelfs.
6. Melting of land-supported ice would cause sea levels to rise but cannot happen without volcanoes.
7. East Antarctica has been gaining ice in spite of 400 ppm CO_2 concentrations.
   East Antarctica does not have a source of heat below the ice to cause melting.
   This vast area is very unlikely to begin melting any time soon.
8. West Antarctica's chain of magma and volcanoes along the shore causes floating ice to melt, but will not cause sea levels to rise.
   The amount that melts will be determined by magma and volcanic activity, not CO_2 concentration.
9. Since only offshore ice is affected by the magma and volcanoes, there will not be enough melted ice to raise sea levels significantly.
10. Further, the net change appears to be an increase in ice.
    Ice increases in East Antarctica has been offsetting any loss in West Antarctica.

Ref. #20, Page 40, Is a table of 37 Volcanoes in West Antarctica, known before 2017.
A 2017 study claimed to have found 138 Volcanoes of which 91 were previously unknown.

POLAR BEARS IN THE HOT TUB

Name	Elevation		Location	Last eruption
	meters	feet	Coordinates	
Mount Andrus	2978	9770	75°48′S 132°20′W	Unknown
Argo Point	360	1180	66°15′S 60°55′W	1.4-0.9 million years ago
Mount Berlin	3478	11,411	76°03′S 136°00′W	8350 BCE
Mount Bird	1800	5900	77°16′S 166°44′E	3.8-4.6 million years ago
Bridgeman Island	240	787	62°03′S 56°45′W	-
Brown Peak (Sturge Island)	1524	5000	67°26′S 164°46′E	Unknown
Coulman Island	1998	6553	73°30′S 169°36′E	-
Deception Island	576	1890	62°58′S 60°39′W	1970
Mount Discovery	2681	8796	78°18′S 165°00′E	Unknown
Mount Erebus	3794	12,448	77°32′S 167°17′E	2015
Mount Frakes	3654	11,998	76°48′S 117°42′W	Unknown
Gaussberg	370	1213	66°48′S 89°11′E	Unknown
Mount Hampton	3323	10,902	76°30′S 126°00′W	Unknown
Mount Harcourt	1571	5153	72°24′S 170°6′E	-
Hudson Mountains	749	2457	74°19.8′S 99°25.2′W	210 BCE
Mount Melbourne	2732	8963	74°21′S 164°42′E	1892 ± 30 years
Mount Morning	2723	8934	78°30′S 163°30′E	Unknown
Mount Moulton	3078	10,098	76°06′S 135°00′W	-
Mount Murphy	2703	8868	75°18′S 100°45′W	Late Miocene
Mount Overlord	3395	11,142	73°12′S 164°36′E	-
Paulet Island	353	1158	63°34.8′S 55°46.2′W	Unknown
Penguin Island	180	591	62°06′S 57°55.8′W	1905
Lars Christensen Peak	1755	5758	68°51′S 90°34.8′W	Holocene
The Pleiades	3040	9974	72°40.2′S 165°30′E	1050 BC ± 1000 years
Royal Society Range	3000	9842	78°15′S 163°36′E	Holocene
Seal Nunataks	368	1207	65°1.8′S 60°03′W	Unknown
Mount Sidley	4181-4285	13,717-14,058	77°06′S 126°06′W	-
Mount Siple	3110	10,203	73°26′S 126°40′W	Holocene
Mount Steere	3558	11,673	76°42′S 117°48′W	Unknown
Mount Takahe	3460	11,352	76°16.8′S 112°04.8′W	5550 BC
Mount Terra Nova	2130	6988	77°31′S 167°57′W	Unknown
Mount Terror	3230	10,597	77°31′S 168°32′E	-
Toney Mountain	3595	11,795	75°48′S 115°49.8′W	Holocene
Mount Dimitra	2987	9797	73°27′S 164°34.8′E	Holocene
Mount Brown[1]	1982	6503	76°49.8′S 163°00′E	Holocene
Mount Christos	-	-	56°15′S 72°10.2′W	1876
Mount Waesche	3292	10,801	77°10.2′S 126°52.8′W	Holocene

POLAR BEARS IN THE HOT TUB

CHAPTER TWO
POLAR BEARS - FUEL TYPES AND CO2 CONCENTRATION

In Chapter One we asked about Polar Bears and hot tubs at the North Pole.
We also spent time learning about ice at the South Pole .
NOAA data showed that CO2 has no observable connection to melting ice.
Nor does data show that air temperature is related to melting ice.
We did learn that rising magma has a lot to do with melting Polar ice.

There is much to observe to finally understand why CO2 is increasing.
Chapter 2 is devoted to this understanding.

There are different questions to ask.
Did burning coal cause the sudden 10 fold rate increase in CO2 after 1980?
 CO2 rate of increase jumped from 0.28 ppm/year to 2.6 ppm/year; a 10 fold jump.
Why did the rapid rise start in 1980 rather than 100 years earlier in 1880 with the industrial revolution and increased burning of coal?
Is CO2 concentration in the air entirely about the amount of coal and hydrocarbons burned? What could cause the ten fold jump in use?
Can we economically reduce CO2 concentration?
Governments are spending immense sums to reduce CO2 emissions.
What if that's not the cause of global warming?
What if we are trying to solve a non-problem?

Who has not been to a children's party where helium filled balloons are provided? Children hold onto the string but any balloon that gets away quickly rises to the ceiling and stays there. With no means or method to bring them down to the floor, or eliminate them, they remain for days.

So it is with CO2 in our world. Gas hydrocarbon fuels are burned, producing heat and power. The exhaust gas is lighter than air. Like the helium balloons, the exhaust gas rises above the clouds where there are no chemical or physical means to remove it. Mixing with air cannot create a mixture heavier than air; the exhaust gas continues to rise and mix with air. It does not return quickly to earth where trees and oceans remove CO2 but air circulation gradually returns the CO2 to ground level. Slightly higher concentrations of CO2 allow trees and food crops absorb the CO2 and to grow faster. The table on the next page, Ref. #21, compares the density of exhaust gas from burning of different fuels; Wood, Coal, 'water-gas', Natural Gas and Hydrocarbon oils.

POLAR BEARS IN THE HOT TUB

FUEL TYPE >>>>>>>>		**WOOD**	**COAL**	**WATER GAS**	**NATURAL GAS**	**Hydrocarbon OIL**
Composition of Fuel		$C_6H_{10}O_5$	C	$CO + H_2$	CH_4	($-CH_2-$)
Components of Exhaust gas		$CO_2 + H_2O + N_2$	$CO_2 + N_2$	$CO_2 + H_2O + N_2$	$CO_2 + 2 H_2O + N_2$	$CO_2 + H_2O + N_2$
O2 Required		6	1	1	2	1.5
EXHAUST Components	MW					
N2	28	24	4	4	8	6
CO	28		0	0	0	
CO2	44	6	1	1	1	1
H2	2	0	0	0	0	0
H2O	18	5	0	1.5	2	1
Moles of exhaust gas		35	5	6.5	11	8
WEIGHT		1026	156	183	304	230
Mole Wt of Exhaust gas		29.3	31.2	28.2	27.6	28.8
Mole Wt of Moist air		28.69	28.69	28.69	28.69	28.69
Specific Gravity vs Air		1.02	1.09	0.98	0.96	1.003
RISE OR SINK ??		Slow SINK	Rapid SINK	Rapid RISE	Very Rapid RISE	Very Slow SINK

POLAR BEARS IN THE HOT TUB

Natural gas is the most used fuel now and more and more is burned each day. Note that the specific gravity of exhaust gas from burning natural gas is lighter than air and it must rise in the atmosphere.

CO_2 is also necessary for marine life; sea shells, coral, back bones for fish and protein for dinner tables. They also grow more rapidly. This is good.

In addition to gas, liquid fuels and wood are burned and form CO_2 but less water vapor than natural gas. Burning wood and petroleum creates an exhaust gas which is very slightly heavier than air. This exhaust rises while hot, but when it cools, it settles slowly.
Coal, in contrast, produces a gas that is heavier than air; it does not rise above the clouds. Instead it sinks to the ground and is absorbed by water or removed by trees and plants.

The tall chimneys at coal plants confirm the high density of the exhaust. Chimneys for coal plants are between 800 and 1200 feet tall. They are tall to allow time for dispersion into air before the dense gas reaches ground level. Gas in the chimneys approaches 20 % CO_2 and little oxygen; not something we want to inhale. Computer programs have been developed to predict where the plume will reach earth and the concentration. City and other local laws specify the maximum concentration allowed at ground level and determine the height of the stack. While expensive, they are necessary and required.

In contrast, power plants burning natural gas create an exhaust gas which is lighter than air. They do not have tall stacks but rather one hundred or so feet high are the rule. This also supports the data on exhaust gas density; a low density exhaust gas is created when natural gas is burned whether it be a power plant, the heating system for a large building like a hotel or the small water heater in a house or the kitchen stove.

POLAR BEARS IN THE HOT TUB

An important feature of the "lighter than air stack gas" is dispersion. Do these light stack gases rise or are they dispersed near the ground and disappear?

We can observe birds that soar, like eagles and hawks; they rise on columns of rising air called thermals.
Or we can learn from people who fly gliders and sail planes.
They have learned to seek these rising columns of air and light gases and can soar to heights of thousands of feet, then glide to another thermal.
Using this technique they can travel hundreds of miles.

Depiction of Thermals

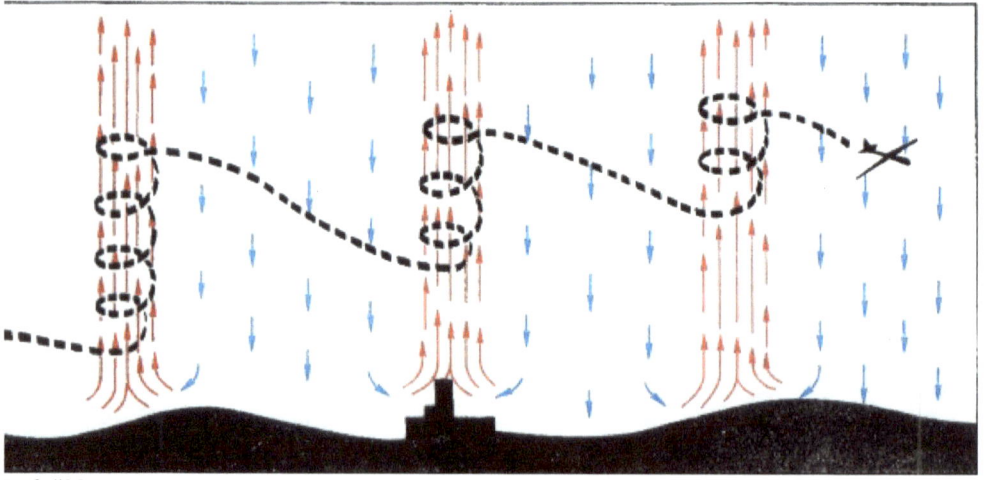

Ref. #22

Two examples of thermals are shown in the above depiction.
On the left, a small area of hot surface on the ground such as a hillside facing the sun. Another, in the center, is a power plant, or it could be a city which is warmer than surrounding areas. The third column on the right is used to demonstrate how the pilot of a sail plane will continue looking for rising thermal to continue on a journey.
Thermals generally occur over a small portion of an area while downdrafts have much larger areas. Updrafts in the thermals are considerably stronger than the downdrafts.

POLAR BEARS IN THE HOT TUB

Sailplane pilots gain altitude in thermals and hold altitude loss in downdrafts to a minimum.

Heating by the sun is the most common cause of thermals, although it may be man-made heat sources such as chimneys, factories, and cities. Cool air must sink to allow the warm air to flow upward in thermals. Note in particular that the "thermal" updraft is a slender column of rising air. It does not dissipate quickly but retains a small diameter. <u>Embed the '**slender**' feature in your mind; it is a very important feature</u>. It defines a "thermal".

    Several examples are shown of rising columns of low density air.
    In the narrative below the Figure, heat sources include chimneys, factories and cities.

We are very interested in "Thermals"; they explain the rapid rise in CO2 concentration. Thermals, these columns of rising air, are less dense than surrounding air, rising thousands of feet, and surprise, not dispersing but maintaining a slender waist, like a chimney.

There are many examples of thermals:
A pot of boiling water creates a column of low density water vapor.
Air over a city is warmer and less dense than surrounding air.
    Asphalt roads, roofs and parking lots absorb sunlight, become hot and heat city air.
    Cities use a lot of energy in lights, factory and office equipment, cars and trucks.
Trees are plentiful in most cities removing CO2 and releasing low density water vapor.
    All these reduce density of air over cities.

There is little air turbulence to disperse either a cloud or a rising thermal.
Who has not watched clouds floating undisturbed for hours? Their altitude is determined by the density of their mixture of dense but very small water droplets, and humid air, which equals the density of surrounding air.

It is time for some basic science about Thermals and CO2.
Simple chemistry that you might have learned in high school is useful but not required in this chapter.

We, the people, are burning a lot of 'clean burning natural gas', and burning more every day.

POLAR BEARS IN THE HOT TUB

What happens to the gas that goes up the stack? We need to know to understand CO2, greenhouse gases, and global warming.

The Chemistry: for burning natural gas.
Natural Gas, chemically Methane, is CH4 to a chemist, made of carbon and hydrogen; both react with oxygen.
 CH4 + air (oxygen and nitrogen) => CO2 + 2 H2O (vapor) + four molecules of Nitrogen for each molecule of oxygen.
This mixture results whenever natural gas (methane) is burned.

The Physics:
CO2 is a dense gas, 1.52 times as dense as air and if pure it will 'sink' to the ground like a rock sinks in water.
H2O vapor is a very light gas, only half as dense as air. Water vapor always rises rapidly. It is the white "smoke" seen above cooling towers and many "smoke stacks" of industry; putting harmless water in the air.
Due to the water vapor in Methanes exhaust gas, the mixture of CO2, water vapor and nitrogen is lighter than air; and it rises!
Each power plant that burns natural gas, creates a lot of stack gas that is less dense than air. Gas leaving the power plant is also hotter than air and like a hot air balloon it rises fast.
The rising column of low density exhaust gas creates a "THERMAL".
The column does not dissipate quickly but maintains a small diameter and continues to rise for thousands of feet.

SO, WE HAVE THERMALS, SO WHAT?
WHY IS IT SO IMPORTANT?

THE NEXT STEP. MORE SCIENCE - PHYSICAL SCIENCE.
We need to ask, what happens to the rising column of CO2 & water-vapor?
1 Until the water vapor is condensed or frozen, this "THERMAL" remains less dense than air and keeps rising to very high altitudes.
2 There are no trees, plants, or bodies of water in the upper atmosphere to remove CO2. Therefore it accumulates like the helium party balloons.

POLAR BEARS IN THE HOT TUB

3   In time, at higher altitude, the CO2 disperses into the atmosphere increasing the concentration of CO2. Analysis of samples is accurate.
4   Burning of natural gas each day means the concentration of CO2 must rise. Increased amounts burned each day means it rises faster.

CO2 HAS INCREASED, AND INCREASE IT MUST.
CHEMISTRY IS CORRECT. PHYSICAL LAWS ARE CORRECT.

Remember:
   Melting Arctic Ice is not about us!
   Melting Arctic Ice is not about CO2!
   Melting Arctic Ice is not even about higher temperatures!

There is another source of CO2 which is man made.
It also raises the level of CO2.
The CO2 from this source does not melt Arctic Ice either.
That source is burning jet fuel in airplanes at 40,000 feet.
Jet fuel is a fossil fuel with the group -CH2- linked together in long chains.
When it burns it produces one molecule of CO2 and only one molecule of water vapor in contrast with burning natural gas which creates two.
The density of the mixture is nearly as dense as air and it neither rises nor sinks. It mixes with air at 40,000 feet. There is nothing to remove the CO2 at 40,000 feet so it accumulates and the ppm of CO2 increases. We should not be surprised that jet planes raise the concentration of CO2.

CO2 IS INCREASING AND IT WILL CONTINUE TO INCREASE.
   CHEMISTRY IS CORRECT. PHYSICAL LAWS ARE CORRECT.

We are studying why CO2 has increased. This analogy may help.
Consider all the CO2 generated by burning methane which produces the lighter than air stack gas. It rises into our "tank" of atmospheric air. This tank is very

different than all other tanks; the bottom is spherical - our 8,000 mile diameter earth. There are no sides and no roof; air is contained by gravity.

Any CO_2/H_2O stack gas we generate 'floats' to the top of the tank and remains there until dispersion and currents carry it 5 to 10 miles downward to plants, trees and oceans to remove it. During this time it accumulates and we see concentrations of 400 ppm instead of 280. It can be no other way.

POLAR BEARS IN THE HOT TUB

The following chart has been created using data from NOAA ice charts and data from the Mona Loa Hawaii laboratory. CO_2 concentration and other data is plotted on the left against years on the lower axis.
It is not "smoothed" to create a CO_2 hockey stick curve but uses actual data points for more information.

POLAR BEARS IN THE HOT TUB

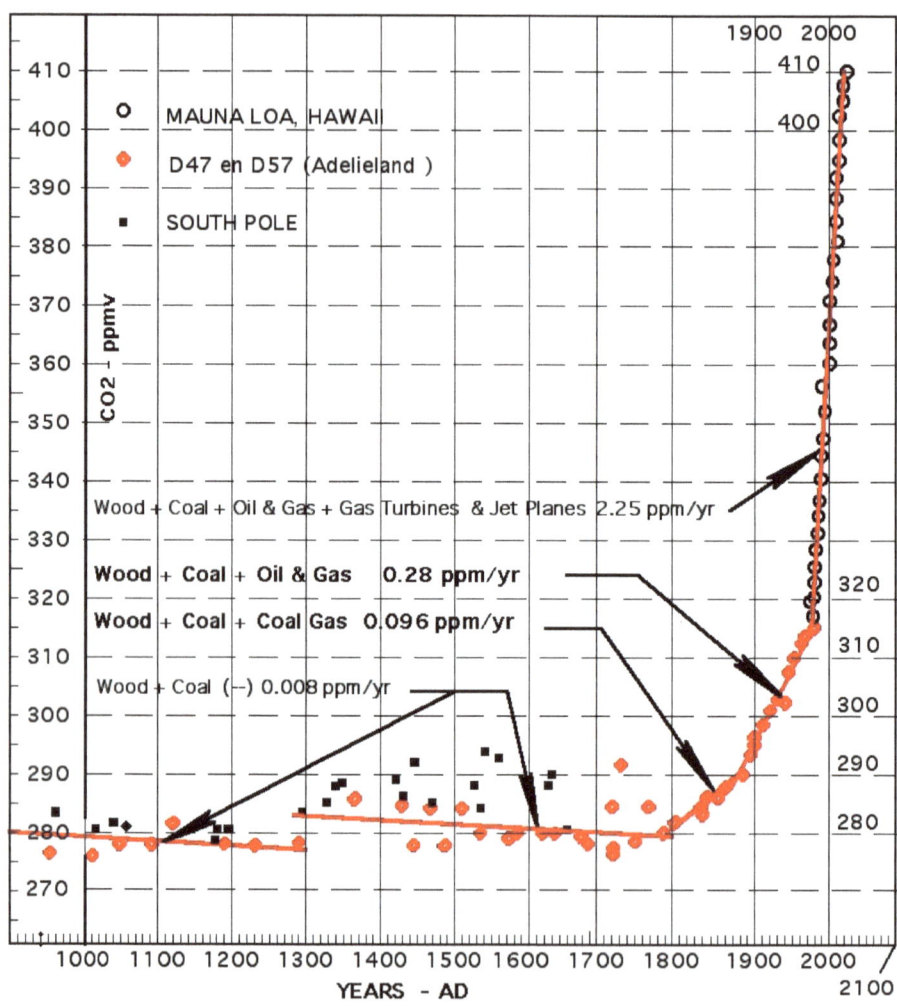

Ref. #23

POLAR BEARS IN THE HOT TUB

The following is informative:
1 Starting over 1000 years ago, CO2 ppm was decreasing 0.008 ppm per year.
2 The eruption of Mount Salamas in 1257 increased CO2 concentration from 280 to about 285 ppm.
3 And CO2 concentration again decreased at about 0.008 ppm per year.
4 Suddenly, about 1780, CO2 started rising at 0.096 ppm per year.
Such a major change from MINUS .008 ppm per year to **plus** demands an answer. Is it a coincidence that Chemistry provides an answer?

Mr Priestly and others discovered in 1774 that "air" was not "air" but rather Oxygen, Nitrogen, water vapor, CO2 and other gases.

Almost immediately, others using this new chemistry, learned that combining white hot coal with water or steam in the absence of air, produced a mixture of gases, called 'water-gas'; a mixture of one Carbon monoxide molecule (CO) and one Hydrogen (H2) molecule. Both burn with oxygen to release heat. It was much easier to use as a fuel than coal. It could be piped to the user rather than carried. A valve controlled energy.

Exhaust gas from burning 'water-gas' is lighter than air; a thermal plume of CO2 and water vapor is created that rises high into the atmosphere. The concentration of CO2 started rising immediately at about 0.1 ppm per year. Extending this rate to 2018 would mean a CO2 concentration of 295 ppm, not 410.

CO2 INCREASED, AND IT CONTINUED TO INCREASE.
CHEMISTRY IS CORRECT. PHYSICAL LAWS ARE CORRECT.

Chemistry began to influence many areas:
Steel makers learned that steel with high carbon content was brittle . Very low carbon content created a malleable steel that was still quite strong. Intermediate levels of carbon content allowed making a malleable yet strong steel. It could be machined into pistons and bolts for steam engines. Large steel plates could be made and rolled and riveted to make boiler drums and pipes.
The industrial revolution had begun.

POLAR BEARS IN THE HOT TUB

HOW DOES BURNING COAL AFFECT CO2 CONCENTRATION?

So much has been published about the dangers of burning coal.
Remember the photos of coal plants in the introduction?
Coal is carbon and chemistry shows us that the gas going up the chimney if only carbon is burned will be CO2 and nitrogen, no water vapor.
This chimney gas is heavier than air and disperses to the ground.

Have you noticed the very tall 'smoke' stacks at a coal burning power plant? They are tall because the dense gas needs to be dispersed before it reaches the ground. Computer programs have been developed to predict concentration of CO2 downwind of the stack; the programs calculate the height required to meet regulations.
Typical heights are 800 to 1200 feet; very costly but necessary to disperse the dense gas.
Trees, plants, oceans and clouds all absorb the concentrated CO2 from the stack gas. Only a portion rises into the atmosphere to increase it's CO2 concentration.

Burning "as mined" coal creates problems because coal is not only carbon. Coal also contains sulfur and nitrogen compounds which react with oxygen and are very harmful in the atmosphere but can be and are being scrubbed successfully from the exhaust gas from many power plants.
  This will be discussed later in Chapter 3

SO CO2 INCREASES; WHAT DOES CO2 ACTUALLY AFFECT?

CO2 has a lot to do with many things because environmentalists and politicians decided:
1 Gas from burning coal is very dangerous; Coal must not be burned.
Many of the world governments have committed to reducing CO2 to 'save the planet'.

POLAR BEARS IN THE HOT TUB

2 Very costly programs are being implemented on a worldwide basis.
3 Cost is not an issue when the planet and all life is at stake.
4 Coal plants and 'dangerous' nuclear plants must be shut down while predicting solar/wind and 'future batteries' will provide cheaper and also reliable power. However there are no backup batteries available in anywhere near sufficient quantities or cost to provide "always on" power for several to many days.
   Therefore more gas plants are being built because natural gas is considered "clean" and there must be power when the sun does not shine.
Until affordable storage is available gas plants are required to provide 3/4 of the total power and to provide power for times when the sun hides.
A hazy sky reduces solar power to 10 to 30 percent, sometimes for days.
No energy, no jobs, no food.
5 Cities in California are forced to embrace planning for 'sustainability'.
   No fossil fuel burning vehicles; power only from solar; dense urban growth and restricted "sprawl'.
6 Many clamor for faster and more drastic action even though there is not enough money for 100% carbon free, or nuclear free power or for:
   Homeless people
   Education which is below the worlds average
   Deteriorating roads and bridges
   Insufficient health care for all citizens
   Medical research.
7 Our US Navy is considering where to build new Navy Ship Yards.
Present sites will be under water when sufficient ice melts.
8 Coastal cities are planning for evacuations of millions of people because the sea level "might" rise and flood the areas.

YES! IT MATTERS A GREAT DEAL THAT WE UNDERSTAND.
Think it through carefully:
   Until there is economical battery power, shutting down coal and nuclear will require more power plants that burn natural gas. This will increase the CO_2 concentration ever more rapidly.
Is this not at cross purposes with the intent to 'save the planet'?

POLAR BEARS IN THE HOT TUB

ANOTHER MAJOR CHANGE IN CO2 CONCENTRATION

For almost 100 years, from 1800 to 1900. CO2 increased at a steady rate of 0.096 ppm per year.
Using more 'water-gas' meant accumulation of CO2 in the atmosphere. Science works!

If wood and coal were the first fuel and combustible 'water-gas' the second, then the third phase began in 1859 when Mr Drake found it was possible to obtain oil and gas by drilling. The first well was in Pennsylvania.
Oil could be separated into fractions of kerosene and oil and 'lighter' liquids and gas fuels. The years of hydrocarbon fuels had begun.

The oil boom was on - and CO2 concentration started rising immediately at 0.281 ppm per year. That is three times as fast as with 'water-gas' and wood and coal. Use of energy from coal and 'water-gas' did not suddenly increase by three times! It was cheaper to drill for gas and also a new type of energy was available. There was kerosene available for lamps and heavy oil for boilers; gas pipes were used for cheaper natural gas. 'Water-gas' disappeared.
In a few years, the automobile burning gasoline became common as well as the locomotive and trucks burning diesel fuel. The industrial revolution continued.
   Does chemistry and physics get involved? Of course.
Fossil liquid fuels are long chains of linked -CH2- units.
The exhaust gas from burning liquid fuel contains one hydrogen molecule for each CO2 molecule; the mixture has a density near the density of air. That means it does not create a 'thermal' but stays near the ground and is dispersed mostly near the ground and accumulates in the air for analysis.
There will be more discussion of liquid fuels in Chapter 4.

ONE MORE INDUSTRIAL REVOLUTION AWAITS DISCUSSION!
   During World War 2, nations determined that enormous horsepower could be generated from light reliable gas turbines. Twenty years later the turbine engine

became the engine of choice. Gas fueled turbines drove generators and turbines fueled by liquid jet fuel powered our planes rather than gasoline piston engines and propellers.

Exhaust from gas turbines burning methane created thermals, rising to high altitudes and jet airplanes burning liquid jet fuel created a 'neutral' density exhaust at 40,000 feet. Both added to the CO2 in our atmosphere and CO2 accumulated with a vengeance.

Suddenly, the rate of increase changed from .28 and to 2.5 ppm per year and more; a ten fold increase.

Please study this graph of NOAA data and each statement.

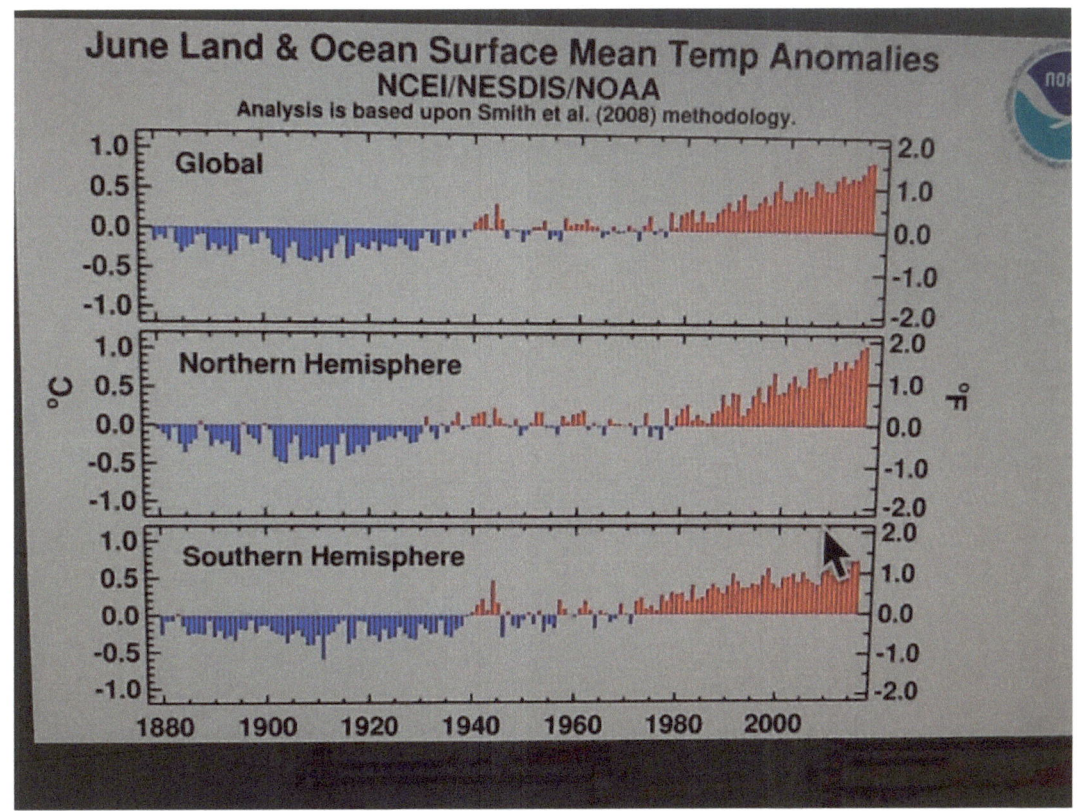

Ref. # 24

POLAR BEARS IN THE HOT TUB

1. Between 1880 and 1980 there was no increase in temperature while CO2 increased from 290.8 ppm to 338.8 ppm.
   In fact, temperature decreased about 1 degree F from 1880 to 1910.
   Then, from 1940 to 1980 there was zero increase in temperature for forty years while CO2 increased to 338.8
2. Population increased with effective medicine (chemistry) and more food production and distribution.
   Population increased while temperature went down between 1880 and 1910, then returned to 1880 levels in 1940, and for 40 years, until 1980 when temperature finally began to increase. With CO2 increasing all this time, there is obviously no connection between CO2 and temperature.

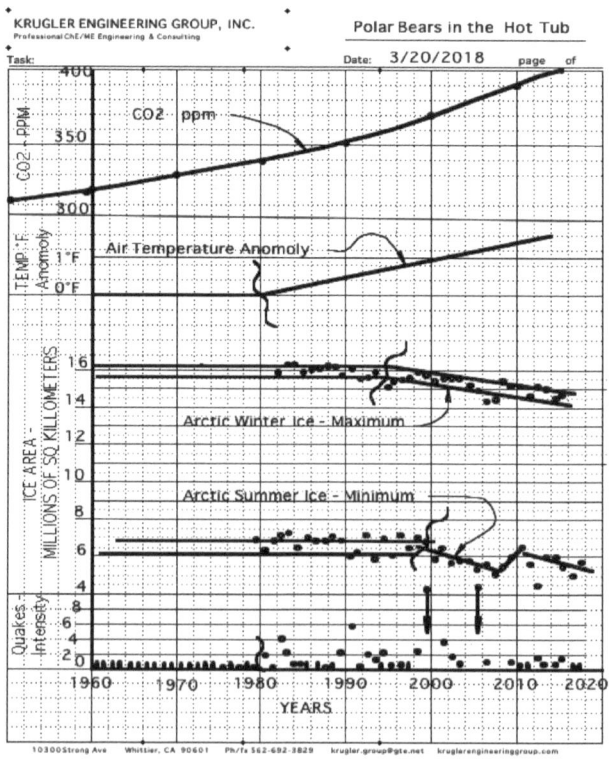

Ref. #25

Note: Dots in the lower graph are a measure of earthquakes and Magma.

POLAR BEARS IN THE HOT TUB

3. Have you noticed that increases in CO_2 actually happened whenever there was a change in the type of fuel available.
4. Population increased with effective medicine (chemistry) and more food production and distribution. There were major increases between 1920 (0.6%/year) and 1960 (2.1%/year) then sudden decline to 1.3%/year currently. The United Nations forecasts further decreases in rate to 0.1% by 2100. That is a rate that barely maintains population.
Population increased rapidly while temperature went down between 1880 and 1940, then returned to neutral for 40 years while population soared, and finally in 1980 CO_2 began to increase while population was decreasing. (See "World Population growth, 1750 to 2100") With CO_2 increasing all this time, there is obviously no connection between CO_2 and population.
5. Melting of Arctic Ice began suddenly in 1995, 100 years after CO_2 was rising.
6. All data is pointing to an unrecognized source of energy which is raising temperature temporarily as it has during each cycle of 110,000 years.

All this is can be seen in this graph of time vs CO_2 and other variables.

There is a another underreported factor which caused the sudden increase in air temperature starting in 1980 and the melting of Greenlands vast fields of ice. It will be discussed in Chapter 5 as "Carbon Soot".
The more we learn the more we discover!

POLAR BEARS IN THE HOT TUB

HOW DOES BURNING COAL AFFECT CO2 CONCENTRATION?

As noted earlier, the very tall 'smoke' stacks at a coal burning power plant are tall because the dense gas needs to be dispersed before it reaches the ground.
Only a portion rises into the atmosphere to increase it's CO2 concentration.

SO CO2 INCREASES; WHAT DOES CO2 ACTUALLY AFFECT?

CO2 has a lot to do with many things because environmentalists and politicians decided:
1 Gas from burning coal is very dangerous; Coal must not be burned.
Many of the world governments have committed to reducing CO2 to 'save the planet' and coal is a major component of the plans.
2 Very costly programs are being implemented on a worldwide basis.
3 Cost is considered to not be an issue when the planet and all life is at stake.
4 Coal plants and 'dangerous' nuclear plants must be shut down while solar/wind and 'future batteries' are expected to provide cheaper and also reliable power. However there are no backup batteries available in anywhere near sufficient quantities or cost to provide "always on" power for several days.
    Therefore more gas plants are being built because natural gas is considered "clean" and there must be power when the sun does not shine.
Until affordable storage is available gas plants are required to provide 3/4 of the total power and to provide power for times when the sun hides.
A hazy sky reduces solar power to 10 to 30 percent, sometimes for days.
No energy, no jobs, no food.
5 Cities in California are forced to embrace planning for 'sustainability'.
        No fossil fuel burning vehicles; power only from solar; dense urban growth and restricted "sprawl'.
6 Many clamor for faster and more drastic action even though there is not
        enough money for 100% carbon free, or nuclear free power or for:
        Homeless people

POLAR BEARS IN THE HOT TUB

    Education which is below the worlds average
    Deteriorating roads and bridges
    Insufficient health care for all citizens
    Medical research.
7 Our US Navy is considering where to build new Navy Ship Yards.
    Present sites will be under water when sufficient ice melts.
8 Coastal cities are planning for evacuations of millions of people because the sea level "might" rise and flood the areas.

YES! IT MATTERS A GREAT DEAL THAT WE UNDERSTAND.
  Think it through carefully:
  Until there is economical battery power, shutting down coal and nuclear will require more power plants that burn natural gas. This will increase the CO_2 concentration ever more rapidly.
  Is this not at cross purposes with the intent to 'save the planet'?

POLAR BEARS IN THE HOT TUB

ANOTHER MAJOR CHANGE IN CO2 CONCENTRATION

For almost 100 years, from 1800 to 1900. CO2 increased at a steady rate of 0.096 ppm per year.
More 'water-gas' use meant accumulation of CO2 in the atmosphere. Science works!

If wood and coal was the first fuel and combustible gas the second, then the third phase began in 1859 when Mr Drake found it was possible to obtain oil and gas by drilling. The first well was in Pennsylvania.
Oil could be separated into fractions of kerosene and oil and 'lighter' liquids and gas fuels. The years of hydrocarbon fuels had begun.

The oil boom was on - and CO2 concentration started rising immediately at 0.281 ppm per year. That is three times as fast as with 'water-gas', wood and coal. Use of energy from coal and 'water-gas' did not suddenly increase by three times! It was cheaper to drill for gas and also a new type of energy was available. There was kerosene available for lamps and heavy oil for boilers; gas pipes were used for cheaper natural gas. "'water-gas'" disappeared.
In a few years, the automobile burning gasoline became common as well as the locomotive and trucks burning diesel fuel. The industrial revolution continued.
   Does chemistry and physics get involved? Of course.
Fossil liquid fuels are long chains of linked -CH2- units.
The exhaust gas from burning liquid fuel contains one hydrogen molecule for each CO2 molecule; the mixture has a density near the density of air. That means it does not create a 'thermal' but stays near the ground and is dispersed mostly near the ground and accumulates in the air for analysis.
There will be more discussion of liquid fuels in Chapter 4.

ONE MORE INDUSTRIAL REVOLUTION AWAITS DISCUSSION!
   During World War II, nations determined that enormous horsepower could be generated from light reliable gas turbines. Twenty years later the turbine engine

became the engine of choice. Gas fueled turbines drove generators and turbines fueled by liquid jet fuel powered our planes rather than gasoline piston engines and propellers.

Exhaust from gas turbines burning methane created thermals, rising to high altitudes and jet airplanes burning liquid jet fuel created a 'neutral' density exhaust at 40,000 feet. Both added to the CO_2 in our atmosphere and CO_2 accumulated with a vengeance.

Suddenly, the rate of increase changed from .28 and to 2.5 ppm per year and more; a ten fold increase, and a 26 times increase from burning coal.

POLAR BEARS IN THE HOT TUB

Please study the graph of NOAA below and each statement.

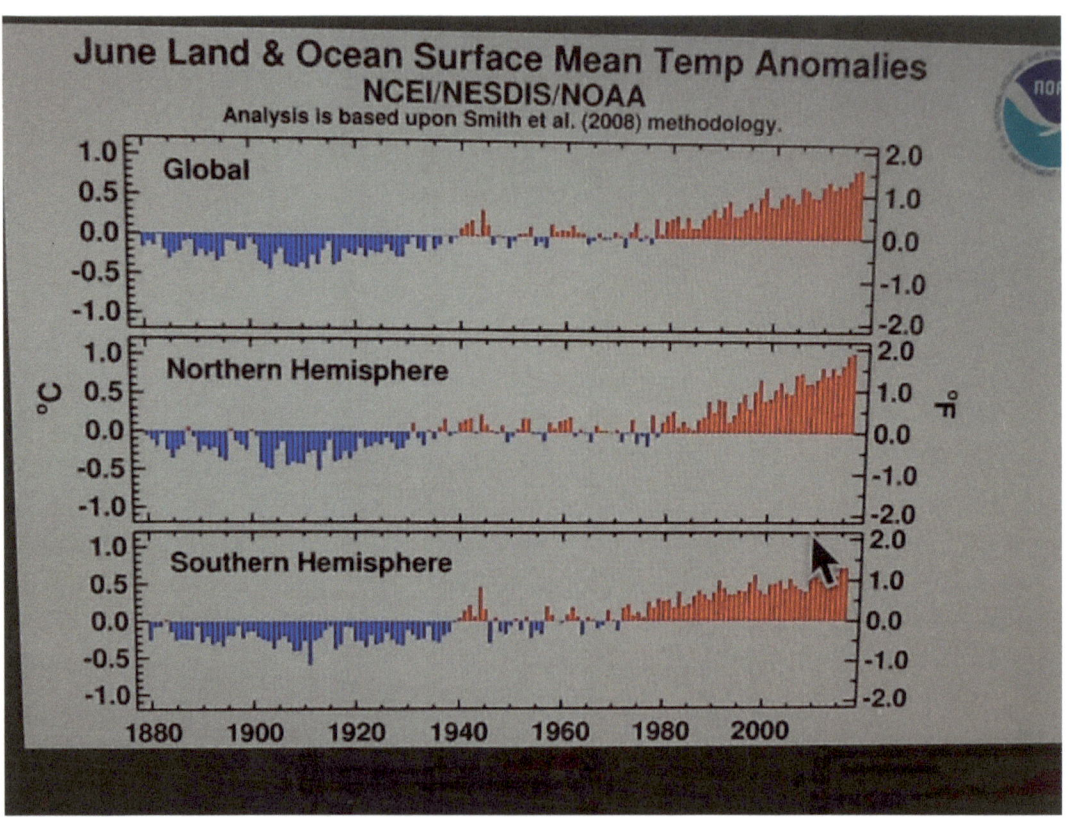

Ref. #26

1. Between 1880 and 1980 there was no increase in temperature while CO2 increased from 290.8 ppm to 338.8 ppm.
   In fact, temperature decreased about 1 degree F from 1880 to 1910. Then, from 1940 to 1980 there was zero increase in temperature for forty years while CO2 increased to 338.8
2. Population increased with effective medicine (chemistry) and more food production and distribution.
   Population increased while temperature went down between 1880 and 1910, then returned to 1880 levels in 1940, and for 40 years, until 1980

when temperature finally began to increase. With CO2 increasing all this time, there is obviously no connection between CO2 and temperature.

There is a another underreported factor which caused the sudden increase in air temperature starting in 1980 and the melting of Greenlands vast fields of ice. It will be discussed in Chapter 5 as "Carbon Soot".
The more we learn the more we discover!

SO! 'SCIENCE' HAS PROVEN THAT THE INDUSTRIAL REVOLUTION INCREASED CO2. THAT IS CORRECT!
HAS 'SCIENCE' PROVEN THAT CO2 HAS INCREASED TEMPERATURE? THE ANSWER IS -- NO!
HAS 'SCIENCE' PROVEN THAT CO2 IS MELTING POLAR ICE AND GLACIERS? NO!
HAS 'SCIENCE' PROVEN THAT POPULATION GROWTH RAISED TEMPERATURE? NO!

IN FACT SCIENCE PROVES THAT CO2 HAD NO MEASURABLE EFFECT!

POLAR BEARS IN THE HOT TUB

CHAPTER THREE
Coal, How Does Burning Coal Affect CO2?

Many questions have been asked in Chapter 1 and Chapter 2.
Questions are essential for learning; lets continue asking.

What if burning the carbon in coal is not a problem for the bears?
Or for Us?
Why was earth's atmosphere no warmer in 1980 than in 1880, after 100 years of burning coal?
Why was the Arctic Ice not affected for those 100 years?
What if, after all the research, and alarm, it is really not about coal?

Polar bears deceive us; so does coal.
Polar Bears are White! We believe this.
Their fur, which we see, is indeed white.
Under the fur is a black skin and a very black bear.

We see coal, and it is black.
Smoke from coal power plants is black; we believe it.
There are many photos of coal smoke, close to 90% show belching black smoke from the tall chimneys.
Yet, there is probably not one smoking stack to be found in the entire United States.

Coal, **we are told,** when burned, causes major problems,
Serious pollution - which causes thousands of early deaths.
- Rising CO2 levels, which causes
- Global Warming, which causes
- Melting ice, which causes
- Sea levels to rise, which causes
- 40 million people living near oceans to be at risk of flooding
- and which raises acid levels in lakes and oceans, which
- destroys coral reefs and marine life.

Actually there are even more questions and many answers to coal combustion.
We are being deceived very badly by 'science' and the media.
Can coal be burned without these problems?

POLAR BEARS IN THE HOT TUB

We have already realized that coal can be burned without the black smoke. It is happening in the USA, and black smoke can also be eliminated from stacks in India and China.

We have demonstrated over a large area that burning coal need not endanger lakes. If we take time to ask, we can learn much more about the problems and solutions.

Coal, which is largely carbon, unfortunately, is not entirely carbon.
There are sulfur and nitrogen compounds demanding our attention.
There are also organic compounds in coal containing sulfur, nitrogen and more.
In short, coal is a mess requiring considerable thought before burning it or banning it.

The carbon in coal, plus air (O_2 and N_2) => CO_2 plus nitrogen, but no extra water vapor.
The density of coal stack gas is 1.09 times as dense as air and it settles to the ground.
It can rise if wet coal is burned, but this is not economical so coal is dried.
It can also rise while hotter than air by about 120 degrees F but cools quickly and descends.

Black smoke was indeed a problem but no longer. Combustion is controlled much better now with accurate oxygen sensors.
The smoking stack is history in our USA, surviving only in photo archives. Precise control of excess air, and Low NOx burner designs eliminates the smoke, reduces NOx and improves efficiency.
There are electrostatic precipitators to remove smaller smoke particles and reduce haze and finally cloth filters in bag houses.
Sulfur and Nitrogen acid gases are removed in wet scrubbers.
We are being deceived very badly by 'science' and the media.

The stacks are 800 to 1200 feet high and very expensive.
Investors do not spend huge amounts of money if it is not required.
Yes, the stack gas does indeed sink to the ground and tall stacks are necessary to disperse it. No one wants to breath stack gas with 200,000 ppm CO_2 but little oxygen, even without sulfuric and nitric acids.
The stacks are still necessary for us, not for plants or lakes.
There are computer programs to calculate how tall a stack must be.
Taller stacks mean lower concentration on the ground near the plant.
Government rules specify the maximum downwind concentration and thus height.
Flowing along the ground, dispersed CO_2 is absorbed by trees, plants and water.

POLAR BEARS IN THE HOT TUB

Is ground level CO_2 bad for our planet? (without SOx and NOx)
No and it is actually helpful.

Plants at the bottom of the food chain require CO_2.
All sea food and coral depend on CO_2. No CO_2, no shells or bones, no sea food, no protein for many Asian nations.
Agronomists have found plants grow faster and need less water with CO_2 at 1500 ppm.
There are coal burning plants near cities and they coexist quite well;
 trees in the city are green and residents are not dying at high rates.

Is there still something wrong with burning raw coal? YES!!
Sulfur, Nitrogen and organic compounds are serious and dangerous pollutants.
Sulfur reacts with oxygen to form SO_2 and SO_3, which with water, form sulfuric acids.
Nitrogen reacts with oxygen to form NO_2, NO_3 etc; which with water forms nitric acids.
These are very strong acids that dissolve in water and destroy lakes and affect oceans.

Has science solved this problem?
Yes! We know how and have done it for many years.
We in the United States have already corrected this problem for the New England States!

Lakes in our New England area were turning acidic and dying from "acid rain" during the 60's. The problem was traced to sulfur and nitrogen in coal used in power plants in Ohio and Pennsylvania.
"Wet scrubbers" were developed and installed to remove the acid gases from coal plant stack gas.
Calcium hydroxide reacts with the gases to form gypsum for wall board manufacture.
Today, over 50% of gypsum wallboard used in construction is obtained from coal plants.
And, the lakes of New England have recovered; actually surprisingly quickly.
Some CO_2 is also removed by the scrubbers but removal is not necessary to create 'clean' exhaust gas free of strong acids, or to protect the lakes and rivers.
These lakes, becoming acidic, then recovering, proves that stack gas from coal plants does indeed settle to the ground. The amount dispersing upward to mix with air did not raise either CO_2 or temperature for 100 years of the industrial revolution which is further proof of the small effect CO_2 has on temperature or melting ice.

POLAR BEARS IN THE HOT TUB

 Is it possible the lakes recovered so quickly because of additional CO2 from the power plants? It is definitely possible but if it has been studied, it has not been reported.

A further problem may exist from sulfuric acid absorbed in sea water.
High School Chemistry 101:
SO3 + H2O => H2SO4 , Sulfuric Acid. Oceans have lots of water to create and dissolve the acid.
H2SO4 + NaCl => Na2SO4 + HCl; Ocean water has lots of Sodium Chloride for this reaction.
 Hydrochloric Acid is a strong acid with unusual properties.
HCl + O2 => H20 + Cl2 Oceans have sufficient oxygen for this reaction to form the Chlorine.
Bottom line: Sulfur from burning sulfur in fuels and also from underwater thermal vents => Chlorine.

Chlorine is a very powerful bleach!
It kills microorganisms and bacteria and will bleach many things.
Chlorine is added to our municipal water to kill bacteria and react with organic material to provide safe drinking water.
Is this the cause of the bleached and dying coral reefs?
One Study by RW Macdonald reports that sea water will react with 1.5 mg chlorine per liter.
link; dfo-mpo.gc.ca
Reactions of chlorine with coral could be and should be studied.
We do not need more reports verifying that coral is being affected by something.
Removing sulfur and nitrogen oxides from fuels should reduce the problem.
Obviously, sulfur and sulfides from underwater hydrothermal vents cannot be controlled. Complete removal of Chlorine will be impossible because of underwater thermal vents but historical levels should be acceptable.

Incinerators are an acceptable method to convert organic compounds to CO2 and water at elevated flame temperatures with sufficient reaction time and adequate mixing. Krugler studied two incinerators that were not effective at a Chemical Plant in California Changing location and method of introducing the foul gas, raising the temperature and providing sufficient time eliminated the problem. Krugler's data was used by the local Air Pollution Control District to write their AP-42 pollution control manual.
   Note: Incineration will not convert sulfur or nitrogen compounds to something safe; they will become sulfuric and nitric acids.

POLAR BEARS IN THE HOT TUB

POLAR BEARS IN THE HOT TUB

CHAPTER FOUR
MORE FOSSIL FUELS - LIQUID HYDROCARBONS

What role did liquid fuels play in the very sudden rise in CO2 concentration after 1880?
Was it also a coincidence that oil drilling provided gas and liquid fuels in 1875 at the exact time that CO2 concentration started increasing from 0.096 ppm/year to 0.281 ppm per year?
Was it also a coincidence that turbine engines for jet airplanes became available in 1958, at the exact time that CO2 concentrations started rising almost ten times as fast at 2.5 ppm per year?

Are the near infinite number of studies and news articles completely accurate in blaming coal?

There is a lot of data causing us to think otherwise.

Notes:
1 Between 1952 and 1963 ice area increased along Russia and Western Greenland in spite of increasing CO2.
2 Between 1963 and 1996, ice area increased all the way to the shore of Russia and to the shores of Canada and Alaska in spite of constantly rising CO2 and air temperatures rising after 1980. There should instead have been a decrease if CO2 and temperature increases cause the ice to melt as models predict.
3 During the 1963 to 1996 period, photos show ice melted along the western shore of Iceland. This coincided with magma activity under the ocean west of Iceland; a clear source of energy to melt the ice.
4 Between 1996 and 2012 photos show there was major reduction in the area of Arctic sea ice. Earthquake activity changed from near zero events before 1980 to numerous yearly events until 2016. Earthquake activity is caused by rising magma, resulting in spreading of the seabed, undersea volcanoes, and thermal vents.
5 Earthquakes foretelling magma appeared off the Western shore of Greenland at a sea-ridge at this exact time. Neither Hawaii's current volcano nor this record of earthquakes can be ignored.

POLAR BEARS IN THE HOT TUB

This history is similar to the decrease in ice 100 years ago, before 1922 as documented by Dr Hoel of Norway. His 1922 report is included at the end of this chapter. After his report, ice increased to full ice cover in 75 years as shown in the 1996 map. How would it be possible for a uniform blanket of CO2 enhanced air to heat the volcanic island chain north of Iceland by 15 degrees between 1960 and 2015 but no other place on earth by near that amount?

EARTH'S WIND PATTERNS

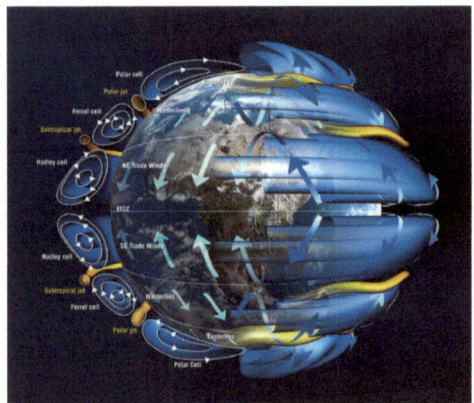

Ref. #28

Chapter Two introduced "thermals" and compared the warm air that rises as a "thermal" to low density stack gas generated by burning methane. Once in the upper atmosphere, stack gas is mixed with air increasing the CO2 concentration. It is removed only at earths surface and slower than it is added.

Another concept is our atmosphere is 'contained' as in a large storage tank.
Our 'earth tank' is very unique; the only one on earth.
This 'earth tank' has a bottom that is not flat but is spherical; the earth itself.
This 'earth tank' has no walls and no top. Air is "contained" above the earth.
It is similar to a cylindrical tank in that a 'light' gas added at the top of a tank at 40,000 feet, does not rush to the bottom where plants and water would remove it. Instead it slowly mixes with air and thus is removed very slowly.
When thinking of turbulence in the sky, remember that clouds remain for hours without change in size, shape, or elevation as they drift with the wind.

POLAR BEARS IN THE HOT TUB

This Chapter compares liquid fuels to natural gas and to coal in their impact on CO_2 concentration and global warming.
These include, Jet fuel, diesel, kerosene, gasoline, and heavy fuel oil.
In other words, Planes, Trains, Trucks, Automobiles and Ships.

Early Trains burned coal - CO_2 exhaust was more dense than air and stayed close to the ground and did not mix into the higher atmosphere to any extent.
CO_2 continued to decrease as it had before the "industrial revolution" began.
Modern Trains - burning diesel fuel - generate an exhaust that rises when it is warmer than air and then slowly sinks.
Ships - Marine cargo ships often burn a very heavy fuel oil and the exhaust is also heavier than air especially if sulfur content is high.
Jet Planes - burning jet fuel - have exhaust that is also neutral density but is released at 40,000 feet; it mixes with air and is not removed. The result is CO_2 ppm is increased by burning jet fuel as well as burning methane and CO_2 ppm is increasing rapidly.

Density of exhaust gas from burning any fuel gas is less dense than air; about 85 percent as dense as air.
Remember that stack gas from burning coal settles to the ground and does not accumulate. It is the only fuel that creates a dense exhaust. See table in Chapter one.

All affect the atmosphere; the question is, how much? Jet planes are part of the abrupt spike in atmospheric CO_2 which started in 1980 when jet engines changed air travel. At the same time, gas fueled turbines started driving generators in power plants, rather than additional coal plants.

Rising CO_2 levels caused by burning hydrocarbons does not
increase temperature as was discussed in Chapter One. However, there are two serious problems from jet fuels; one, jet fuels are not yet treated to remove sulfur and nitrogen, and two, jet engine exhaust contains an infinite number of carbon nanoparticles.

Sulfur, which if not removed, causes aerosols and acid rain and is very harmful.
The second is the soot created when fuels are burned.

Impact of the problems caused by jet fuels CO_2 and particulates cannot be understood without understanding circulation of air around our earth. Diagrams are provided to help.
The globe on the left, considers only surface wind and is incomplete, even misleading.

POLAR BEARS IN THE HOT TUB

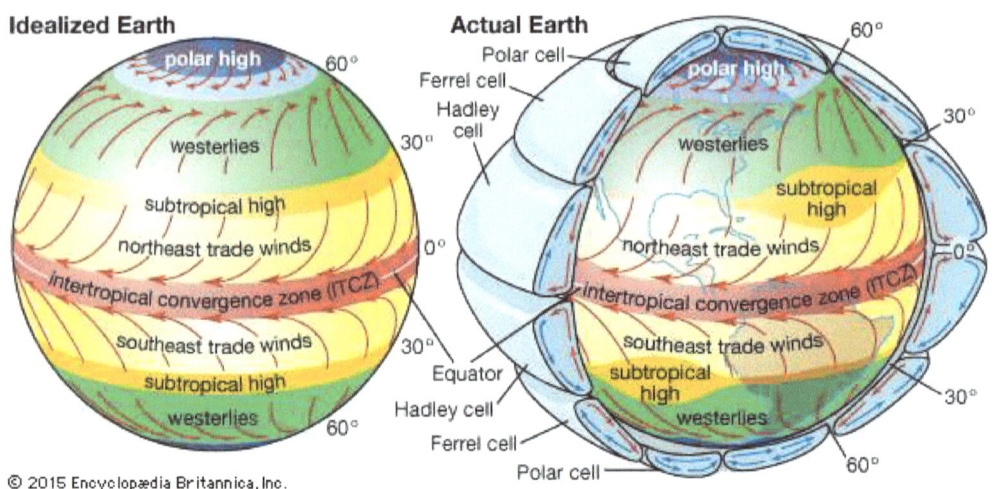

Of the two globes, the globe on the right includes air circulation at higher altitudes. It shows the location of the Hadley cell, the Ferrel cell and the Polar cell. Surface winds are also shown which is important to understanding our weather.

Neither diagram shows air currents from the Hadley cell that flow above the Ferrel cell toward the poles. This is critical to understand facts that appear on the earth surface. Both show the surface winds. Note that the Westerlies flow toward the North Pole but also eastward. Hurricanes in the United States flow within the Ferrel Cell. They originate in the area between the Hadley and Ferrel Cells, make landfall on the southeast cost. Then they follow the Westerlies North and East over the New England states.

The diagrams are very helpful but the opposite page diagram shows the flows more clearly including the particulates.

Starting with the trade winds which flow toward the equator, moist, warm air at the equator rises and returns toward the poles. A large portion circulates in a Hadley Cell, but some flows over the top of the circulating Farrell cell and joins the Polar Cell.

Jet planes fly at 40,000 feet and their exhaust with sulfur acids and soot is carried northward to the Arctic Circle and to all areas in between. Embed this fact in your mind. It is critical to understand many environmental problems from burning fuels containing Sulfur and Nitrogen to the problem with soot.

POLAR BEARS IN THE HOT TUB

The United States has a history with the damage of sulfur and nitrogen and also the solution. Coal plants need to always use alkaline scrubbers to remove the acid gases.

Acid scrubbing been done effectively in the Eastern United States. There was a time in the 1960's when lakes in the Northeast were turning acidic and dying. "Acid Rain" became a national concern. Legislation required power plants to install scrubbers. The lakes recovered surprisingly quickly. The scrubbers produce gypsum which supplies much of the drywall for construction. Sulfur is removed from the environment permanently. Remember, it is not the CO2 from the plants burning coal, it is the sulfur and nitrogen which can and is being removed.

A further probable problem with sulfuric acid in lakes and oceans is it reacts with sodium chloride to create Chlorine gas. Chlorine gas is soluble in water and is a powerful bleach and may cause coral bleaching and dying. Our municipal drinking water is rendered free of infectious organic organisms with levels as low as 2 ppm. Industrial cooling towers prevent algae growth with 10 to 20 ppm. We would be wise to explore the impact of low levels of chlorine on both coral and the ocean food chain.

Sulfur content has not been controlled in jet fuel and often the sulfur content exceeds 2500 ppm. The ASTM global guideline suggests a maximum of 3000 ppm. For contrast, diesel fuel in California must be below 15 ppm. All SO2 from burning sulfur winds up somewhere on earth, and since oceans comprise 71% of the surface, over 70 percent is likely absorbed in the oceans.

Airlines can help reduce these problems without government help by simply refusing to buy any fuel with over 15 ppm sulfur. Refineries will charge more money per gallon but will quickly adapt. This would reduce sulfate particles, acid rain and possibly even the bleaching and death of coral and sea life.

There is however another major problem that will be discussed in detail in chapter five. Carbon does not burn completely in any of these engines but very, very small particles of carbon as well as CO2 are formed.

Turbine engines in jet planes eject an amazing number of carbon particles.

One study on military jets found 2 to 3 million particles in each cubic centimeter of exhaust. A cubic centimeter is about the volume of the end of our little finger beneath the nail.

Carbon particles in the air are indeed a serious problem deserving a separate chapter in this book.

The USA EPA agreed; their report, "Report to Congress on Black Carbon" in 2010 ran to 388 pages. This report discusses many sources and many areas impacted by black carbon but does not include jet engines in its summary. Jet planes had been flying for over 50 years from 1958 to 2010. Jet travel had been well established; carbon from turbine engines deserved greater recognition. There is still very little research on how to reduce the carbon.

Chapter five will provide basic information.

Chapter five will discuss formation of the carbon particles and their impact.
- Carbon particles absorb sunlight, are heated and heat air.
- Carbon particles settle on ice and absorb sunlight and melt ice .
- Carbon particles settling on earth increase the absorption of sunlight.
- Carbon particles settling on water will also increase sunlight absorption.

CHAPTER 6 --   The future of the Polar Bears - Our Future

WE HAVE CHOICES TO MAKE

CHAPTER FIVE
Greenland's Black Ice

Jet Planes were very exotic in 1950 and still glamorous in the 60's. Propeller planes were already using turbine engines rather than piston engines.
 I moved my family in 1966 to live under the jet plane approach to Los Angeles Airport for over 50 years. We often watched in awe as planes from the east flew overhead with flaps and wheels down and engines whining under high power as they flew on to land at LAX 20 miles away. Others, arriving from the west made U-turns overhead to enter the flight path. Their elevation at our house was probably 2,000 to 4,000 feet above us.

There was a darker side to the glamour however. A white paper towel would turn black when wiped over a kitchen counter. Each day, counter tops which appeared clean, were re-coated with black soot which settled from the exhaust. The particles are extremely small - not visible even with the most powerful optical microscopes. A single particle is not visible under an optical microscope and a clump looks like black paint. Powerful electron microscopes are necessary and particles must be pre-coated with a gold film in order to be filmed. I paid two inspection laboratories with electron microscopes, to provide pictures of my samples of soot; both failed. The second one assured me after their attempt failed, that their microscope could photograph particles as small as 2 to 3 nanometers if they prepared the samples by coating them first with gold and then scanning for several hours; the cost would be over a thousand dollars per image with no guarantee for success of any one image. I postponed the opportunity to spend thousands of dollars of savings on the possibility of seeing a photo or two.

With that much soot, there was concerned about the health effects on family and residents inhaling the soot for over 50 years. However, there was nothing notable in the 14 adults or their 15 children who grew up in this immediate area. No asthma, no missed work or missed school days or any illness. No pulmonary issues. Medical reports claiming serious medical problems from fine particulate soot should be scrutinized very carefully for other vectors.

How is this jet engine soot formed?

POLAR BEARS IN THE HOT TUB

Picture in your mind, jet fuel burning in one of the several combustion chambers of a jet engine. It is a microscopic world.

Jet fuel is injected at great pressure, 20,000 to 25,000 psi, through small nozzles to "atomize" the fuel into droplets. Higher pressures create smaller size drops. Size of droplets in current engines is very small, almost like the dust we see floating in a shaft of sunlight in our houses. Each small droplet enters a white hot radiant chamber along with compressed air.

The outer layer of this miniature oil droplet starts to boil immediately - and starts burning - creating an inert shell of CO_2 and water vapor around the droplet. This inert shell is a barrier to oxygen reaching the droplet and impedes further burning. The inert shell becomes thicker with any burning around the ever smaller droplet. It is so small it floats with the current, like the dust in our living room, not burning completely, but cracking into carbon and hydrogen. Hydrogen gas created by the cracking burns readily, but the carbon does not and leaves the engine as an extremely small particle; probably not round and sometimes clustered with other small particles like a small bunch of grapes.

This means that very nearly every droplet of oil creates at least one carbon particle, and most will leave the engine unburned. As stated earlier, 2 million to 3 million particles have been measured in single a cubic centimeter of jet engine exhaust. Jet engine fuel creates an exhaust that is nearly as dense as air due to the water vapor formed from burning hydrogen. It remains in the air at 40,000 foot elevation of cruising jet planes. While carbon particles are heavier than air, they are actually very light and very small. The hot mix of CO_2, water vapor, nitrogen and carbon particles is momentarily lighter than air and floats upward. The study of jet engine exhaust reported soot particle diameters between 2 and 60 nanometers; (equal to 0.000000002 to 0.000000060 inches in diameter). From 40,000 feet, they have 7 to 8 miles to settle before reaching earth. Rather than settle, they flow toward the poles with atmospheric currents at 40,000 feet. Air currents north of the equator, where planes are shown flying in the diagram, must carry much of the soot northward toward Greenland and the North Pole.

What problems does this soot cause? Actually there are two problems!

The first problem. Even while nanoparticles are floating, they are a black solid that absorbs nearly all sunlight that strikes each of the untold millions of particles. Heat transfer dictates that temperature of the particle is increased by sunlight and that air in contact with those particles is also heated. Air temperature of the northern hemisphere where most air travel occurs will be heated and heated more than southern air temperatures.

POLAR BEARS IN THE HOT TUB

Jet engines burn 6,000,000 barrels of oil a day. A barrel of fuel weighs 450 impede; that amounts to 2,700,000,000 lbs of fuel a day (2.7 billion lbs.) and a near infinite number of particles.

POLAR BEARS IN THE HOT TUB

GLOBAL AIR CIRCULATION CELLS

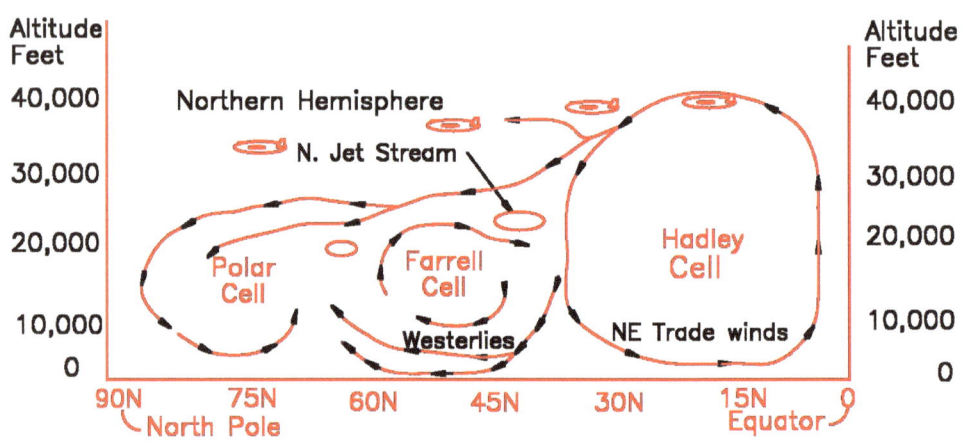

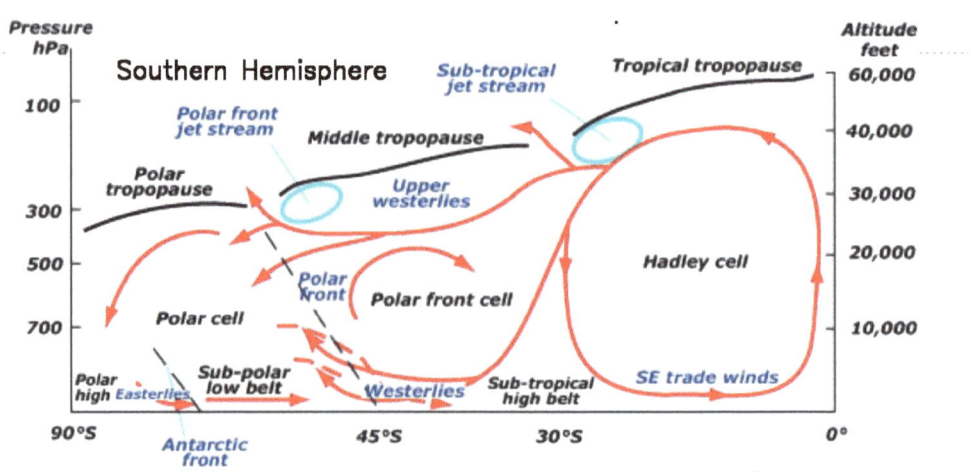

POLAR BEARS IN THE HOT TUB

Scientists have tried to calculate the impact of this carbon on global warming. One four year study is a 232 page report in the Journal of Geophysical Research, January 2013, (e360yale.edu) is summarized by Carl Zimmer, January 17, 2013.

The report estimates that carbon soot can cause more warming than that attributed to the CO_2 blanket; 1.1 watts to 2.1 watts vs the 1.56 watts attributed to CO_2. The maximum of 2.1 watts is over 30% more than the 1.56 watts for CO_2. The report does not mention jet fuel as a source of carbon released at 40,000 feet or the ground sources generating low density exhaust that rises to join the jet exhaust. The extent of this first problem, heating of the atmosphere, is not really understood. More research is required.

The second problem with this soot!
Air currents carry the soot northward to Greenland and the North Pole and then downward to the surface of the ice. Greenland is experiencing many miles of "black ice".

To see many photos of the ice, search "Greenland, Black Ice, photos". One photo is reprinted below. Once on the ice it absorbs more sunlight than soot-free ice and increases melting rate several times.

It would indeed be useful to know where the soot actually comes from so the problem could be solved. The author has not found a chemical analysis of the soot.
Electron microscope photos of nano-particles taken from the ice would be quite expensive. Each electron microscope photo might cost over a thousand dollars but this is a pittance compared with the knowledge developed.
A simple chemical test would show the soot is not from burning wood as has been speculated. Wood ash contains calcium and other metals which are easily detected by a simple chemical analysis. Photographs could be compared with photos of particles from jet engine exhaust.
Instead of an analysis, many millions of dollars are spent measuring the increased absorption of the black ice. Without the soot there would be nothing to measure.

Is the soot caused by burning forest fires?
  Wood smoke from a hot fire is normally white, not black. Further, smoke plumes from the Edmonton, Canada fires were photographed from space; the plume was carried by winds directly south over central states of the US, then East over New England and out to sea toward Scotland. The plume from this fire never passed over even southern Greenland. A statement that the soot is forest fire ash was an uninformed opinion and wrong.

POLAR BEARS IN THE HOT TUB

Photo by Dr. Jason Box - www.jasonbox.net

Are there solutions to the soot problem? Yes!
According to news releases development of even higher pressure pumps and injectors than 25,000 to 35,000 psi can create particles so small that the droplet is entirely burned; no soot is formed. If true, this would be extremely helpful.
Requiring all engines to be equipped with this high pressure system would eliminate the soot and by reducing unburned fuel, improve economics.
This could be accomplished very quickly and should be pursued.

Next, oxygen within a fuel molecule causes formation of CO_2 gas and water vapor inside the burning droplet. The gases expand explosively and break up the droplet for faster burning without coking. This type of molecule exists in biodiesel fuel. Blends of biodiesel and jet-fuel mixtures have been flight tested satisfactorily. Biodiesel has been proven to be fire safe, and cleaner burning. Production of more biodiesel should be encouraged.

POLAR BEARS IN THE HOT TUB

Recent reports of tests with hydrocarbon fuels indicate that straight chain hydrocarbons burn much more soot free than unsaturated and cyclic structures. These fuels can be produced from coal by the Fischer-Tropsch reactions but refiners may also be able to reduce the problem molecules.

There may be mechanical factors as well.
Diesel engines and gasoline engines have made improvements by changing the shape of the combustion chamber and location and methods of fuel entry. Mechanical changes might also improve combustion in turbine engines.

In conclusion, there is much that can be done to reduce soot and reduce melting of ice and also warming of the planet.
While reducing heat from Magma beneath sea ice is beyond human capability, soot can be reduced quickly and relatively inexpensively.
Airline corporations can assume responsibility and lead the world.

CHAPTER SIX will explore the future.

CHAPTER SIX
THE FUTURE with WISDOM

CAN THE FUTURE of EARTH'S CLIMATE BE PREDICTED?
OR THE FUTURE OF THE POLAR BEARS?
OR OUR FUTURE? THE FUTURE OF THE PLANET?

Can we, the people, do anything about that future?

Chapter six is a discussion of the data, understanding it and the impact of our activity on future climate.
With understanding comes the ability to predict trends and outcomes.
As Yogi Berra said, "You can see a lot by observing" and
"It is difficult to make predictions, especially about the future"
Yet there is sufficient data to warrant observation and predictions.

POLAR BEARS AND THEIR FUTURE

DATA
-- Polar bears have always lived north of the Arctic Circle along the shore of the Arctic Ocean.
-- They depend on seals for food.
-- Bears live near shore, either on the ice or on shore.
-- Decreasing arctic ice was reported in 1922 by fishermen & Dr Hoel.
-- Atmospheric air temperature did not start increasing until 1980.
-- The ice recovered and then remained constant until year 2,000.
-- Area of ice on the Arctic Ocean started decreasing in the year 2,000.
---- Volcanic islands north of Iceland increased 15 degrees.
---- Air temperature at the North Pole increased 10 degrees.
---- World average temperature only increased 1.5 degrees F.
---- South Pole air temperature may have increased 1 degree.

OBSERVATIONS -- THE NORTH POLE
-- Earthquakes, indicating rising magma, became active in 1980.
-- Earthquakes were active under the ocean and under Arctic islands.
----Earthquakes have been inactive since 2016; two years.

POLAR BEARS IN THE HOT TUB

-- Ice has increased and decreased for 100 years independent of either CO_2, or temperature.
-- Polar Bear population has increased 6 fold from 5,000 +/- to 30,000 since northern nations banned hunting, even as CO_2 was increasing.
---- A last census of Polar Bears found the population stable at 30,000.
-- Ice area has responded in step with magma activity as indicated by earthquake activity along undersea ridges north of Iceland.
---- Ice has melted along the Russian shore but not along Canada.
-- A viral video of a dying bear did not explain that bears live 15 years and that 2,000 bears must die each year from old age or other causes.

FUTURE of POLAR BEARS

-- Population of Polar Bears has probably reached a stable maximum of 30,000 +/- bears depending on their main food supply, the seals. There has not been any reports of massive deaths in any of the 19 areas of the 2017 census. The correlation of Bear Death with loss of ice area has been demonstrated to not exist.
-- Arctic Ocean Ice melts each summer and refreezes each winter. Neither area nor volume has responded to either CO_2 or air temperature but has responded to earthquake/magma activity. Magma activity in the Arctic Region dropped to near zero in 2016. Summer ice has been recovering and will continue to return if magma does not resume activity. With sunspots at a low level this is unlikely.
-- Other researchers have recently confirmed that, based on polar bear surveys, populations have not decreased due to either global warming or melting ice.
-- We can predict that their future is secure, unless ice becomes so thick from the pending ice age, that seals cannot create holes through the ice for breathing. This would deprive bears of their food source.

POLAR BEARS IN THE HOT TUB

FUTURE OF CO2 CONCENTRATION AND ENERGY

DATA
-- Global warming has been attributed to rising CO2 concentrations.
---- The science of greenhouse blanketing gases predicts it.
---- This science is considered "settled" by many.
---- Rising temperatures would then cause melting of ice and glaciers.
---- Melting of ice would lead to rising sea levels and flooded cities.
---- Extreme weather both hot and cold and violent storms result.
---- The Industrial Revolution which started in 1780 with increased burning of fossil fuels is blamed for the rising CO2 levels

OBSERVATIONS
-- The Industrial Revolution did not begin until chemistry evolved.
---- Air was no longer 'air', but N2, O2, H2O vapor and other gases.
---- Iron became steel with the controlled addition of carbon.
------ Steel could be hard or malleable for knifes and machine tools.
------ Now boilers and steam engines and looms could be made.
-- CO2 did not increase with increased population or the burning of coal to fuel boilers prior to the introduction of
-- CO2 did increase however, after chemists learned that hot coal (carbon) and water would react to form a gas mixture of CO + H2.
---- This combustible gas could be piped to a boiler and burned to produce a lighter than air exhaust gas which rose high in the sky.
---- The result was sudden steady increases in CO2 concentration.

-- Increased use of fossil fuels in piston and turbine engines also increased the amount of C02 and carbon soot in the air.
---- CO2 concentration increased three times as fast after 1875 when drilling provided hydrocarbon liquids and natural gas. Both fuels burn to form a lighter-than-air exhaust which rises rapidly high into the sky.
---- There was a tenfold increase in CO2 concentrations after 1980 when the turbine engine was developed to power jet planes and turbines for power plants. Neither population nor fuel use increased 10 times.

-- Carbon soot also increased in like proportions to CO2 causing soot on Greenland to melt it's ice and also to raise temperature on most of the world.

POLAR BEARS IN THE HOT TUB

---- Developing and installing hydrocarbon-free solar and wind plants has not reduced the rise rate of CO2.
-- Countries that have shut down nuclear and installed solar/wind have increased the amount of gas used. Where gas was not available, use of coal and oil increased.
-- Cost to convert to a solar/wind electric economy will be enormous.
-- Assume a solar friendly area - the southwest USA.
---- Constant power is required to assure food supply and delivery.
---- No power, no jobs, no income.
---- Solar plants generate full power between 10 AM and 4 PM.
---- Power output decreases with seasons; when the sun is low .
---- Hazy skies reduce power to 10 to 30% of normal power.
---- Clouds reduce power output even farther.
---- Panels must be cleaned of dust occasionally to maintain power.
---- Output of current panels does decrease slightly with age.
---- One solar plant will be required for power, 10 AM to 4 PM.
---- After 4 PM that plant generates very low to zero power.
---- A second solar plant and batteries will be required, 4 PM to 10 PM
---- A third solar plant and batteries will be required for 10 PM to 4 AM
---- A fourth solar plant and batteries will be required for 4 to 10 AM.
---- More solar and battery plants will be required to provide for each day of low or no sun.
---- Each additional day of standby power will require another set of four solar-battery plants; one day will need eight solar-battery plants.
---- Two days of cloudy/very low output will require 12 plants.
---- All this instead of two gas turbines and one steam turbine.
---- Gas for fuel is stored underground or oil in a storage tank.
---- Even with wind which is also not dependable, the extra plants are required.

SEA LEVEL RISE

Many scientific articles have been written about the amount of ice that could melt - all based on out-of-control global warming.
Previous chapters have shown that polar ice does not respond to CO2 or to air temperature but rather to water temperature.
And water temperature responds to magma activity and earthquakes beneath the sea.

Since these earthquakes have diminished and sunspots are at a minimum, this source of energy is disappearing.

POLAR BEARS IN THE HOT TUB

Sea levels are thus responding to the pumping of water from underground reservoirs. As these reservoirs are depleted, wells will continue to be drilled deeper and deeper until it is not economical.

Level also responds to expansion of water as magma warms that water. Magma activity at the North Pole has decreased. Activity in other areas needs to be studied. Warm water at the South Pole is melting floating shelf ice. Glaciers along West Antarctic can also melt with heat from the over 100 volcanoes, although nearly all are inactive now. East Antarctica does not have volcanoes or magma. This will not be enough ice to severely increase sea levels.
 We can relax.

OUR AND OUR PLANETS FUTURE
-- Government mandates may continue for a time, to aggressively promote a hydrocarbon free world by mandate and subsidies.
---- The result will be an economic wall that will be encountered.
---- The cost of providing numerous sets of four solar-battery plants to provide reliable power will be prohibitive; blackout even more costly.
---- Standby power from gas fired cogeneration (gas turbine/steam turbine plants) operating at 60 + % efficiency will be necessary.
-- Emerging societies will demand more electricity.
---- More power generation from gas turbines will increase the CO2 concentration but not affect global warming. This was discussed in earlier chapters.
-- If soot from burning gas or jet fuel is not eliminated, global temperatures will continue rising with or without solar.
-- If sulfur and nitrogen are not eliminated from fuels, lakes and oceans will turn more acidic affecting coal and marine life.
-- This includes coral reefs.
---- Changing jet fuel by adding oxygen containing molecules helps.
  Biodiesel is one such; MTBE is another additive.
  100 % biodiesel with current engines does not eliminate soot.
---- Improving combustion chambers and/or injectors is also neaded.
---- When the impact of carbon soot is finally realized this will happen.
-- Increased solar/wind will extend the hydrocarbon reserves and is a desirable endeavor.
-- Developing cheaper nuclear power is also essential to provide large amounts of power for future water desalination and future power.

POLAR BEARS IN THE HOT TUB

-- Fusion power may never happen. Materials do not exist to withstand the high temperature and radiant heat that fusion power would create.
-- Expanded and improved medical care will increase human population further increasing the need for power and energy, including transportation, agriculture and air travel fuels.
Clean, potable water for consumption and irrigation water for food crops is becoming a critical commodity. When water wells get too deep, pumping from underground reservoirs will cease. Any process for desalination of sea water, using the best technology still uses enormous amounts of energy, either thermal or electrical to purify the sea water.
Water run-off from pavement and roofs in cities can be captured and injected into reservoirs but only after expensive purification to remove oils and toxic materials. Putting this water into the oceans via rivers will not be acceptable economically.
Both human health and national economies require a reliable source of both electrical energy and transportation fuels 24 hours a day without interruption or black-outs.
Steam turbines, using energy from the hot exhaust from these gas turbines to generate steam, can be on line in as little as one hour and attain full power in two to three hours. Adding steam turbines to gas turbines, known as "cogeneration plants", can generate over 50% more power with the same amount of gas fuel. This is efficiency personified on a large scale.

There is an economic limit on the percentage of total electrical power that is intermittent as are solar and wind. Storage batteries can provide short bursts of power and will be used. However, providing storage in batteries for days of low or very low solar and/or wind is many more times more expensive than storing fuel in gas wells and keeping gas fueled turbines in standby condition. This is especially true in northern climates. Gas turbines can be on line in several minutes and ramp up to full power in 15 minutes when clouds cover solar. They can then be left running through the night and during low/no solar or wind power.
Steam turbines, using energy from the hot exhaust from these gas turbines to generate steam, can be on line in as little as one hour and attain full power in two to three hours. Adding steam turbines to gas turbines, known as "cogeneration plants" can generate 30% and more power with the same amount of gas fuel. This is efficiency personified on a large scale.

Consider this analogy to understand the low density exhaust gases; releasing balloons continually at a birthday party. If, initially hot air balloons are released; they rise but cool quickly and fall. This is equal to burning coal. Changing to helium filled balloons causes the balloons to rise to the ceiling and remain there until the helium leaks out. This is equivalent to burning hydrocarbon fuels. There are now many balloons to count. This is equal to burning hydrocarbons which generate low density exhaust. The exhaust gases rise high in the sky and since CO2 is removed only at ground level, concentration increases rapidly.

Actions being taken to reduce CO2 will have a minor effect on temperature or ice melt or polar bear population. As discussed in Chapter One, where it was shown that neither ice area nor polar bear population responded to changes in CO2 concentration.

Both air temperature and ice area did respond to earthquake/magma activity. It is possible that magma activity will be understood fairly soon but nothing can be done to control it. The earth will experience another 100,000 year cycle from the present warm period to another ice age. We need to expect it and to act accordingly.

A cooler earth will mean much more snow at the North Pole reaching down into Canada, the US and northern nations. Norway, Sweden, Finland and Russia will be affected severely but so will all of Europe. Thankfully, the cool-down period is very slow; 100,000 years.

Additional snow will require better methods to deal with it in cities.

Additional snow, ice and cold weather will also increase need for electrical power, and gas fuels or coal, especially in Northern states and nations. Overcast skies will reduce efficiency of solar panels requiring more 'standby' electrical power.
Review of nuclear power equipment and regulations must be high on the list. Above ground domes to house reactors should be carefully considered. Domes are expensive and construction time is long.
The cost overruns must be analyzed: Why should nuclear plants be the only type of construction with final costs up to five times estimated costs. Our future world cannot accept this.

Chapter five was devoted to carbon soot which is causing much of the global warming - but so far, has not earned respect from the science community. Carbon soot from both turbine engines and piston engines is in dire need of quick and effective action. Steps can be taken to reduce carbon particulates which will affect both temperature and melting ice. These were discussed in Chapters Two and Five. Incorporating oxygen in fuels can be accomplished more quickly than other changes. Manufacture of MTBE, an ether, from petroleum is well developed. Leakage from fuel tanks, a problem for

POLAR BEARS IN THE HOT TUB

MTBE but not ethanol, for commercial airlines could be monitored in real time and would not be a problem.

Sensors and control of fuel/air mixtures is already in use and cannot be improved farther for any substantial gains.

High pressure injectors which can reduce soot should be investigated and made mandatory as soon as possible.

Forest fires are now putting enormous amounts of CO_2 and ash in the air. Planting more trees will not offset this source of CO_2.
Current methods of putting out fires date back thousands of years.
Assembling thousands of men with hoes is now possible but only after a fire has become great. Air delivery of water and retardant does not happen very quickly either and slows but does not put out the fire.
A rapid response fire engine is needed and can be designed and built.
It should also receive a high priority and financial support.

Severe weather, with heavy snow and rain is to be expected.
How else can the world create its next inevitable ice age?

This dissertation is data driven; not based on opinions.
Finally, please contact the author if there are additional data or calculations which would help with understanding.

THANK YOU !

Table of References

Page	Ref. #	Source
4	1	http://berkeleyearth.org/contact/ Berkeley Earth website grant permission (free of charge) to authors, . . . to reproduce their materials as part of another publication and by additionally providing a link to the Berkeley Earth website.
5	2	https://forum.arctic-sea-ice.net/index.php/topic,1461.msg68601.html#msg68601 Image from open forum blog post by diablobanquisa on Arctic Sea Ice Forum
6	3	http://notrickszone.com/2018/07/10/sea-ice-model-projections-in-a-death-spiral- arctic-ice-volume-holds-steady-for-a-decade/, using data from http:// ocean.dmi.dk/arctic/icethickness/txt/IceVol.txt on Danish Meteorological Institute.
7	4	Author's photo of The GEORGE F. CRAM COMPANY'S IMPERIAL GLOBE. South Pole Area Location of volcanoes added by Author.
8	5	https://www.ncdc.noaa.gov/climate-information (data table) plotted by author.
8	6	https://s3.amazonaws.com/files.technologyreview.com/p/pub/legacy/articlefiles/ climatechart.pdf NASA Goddard Institute for Space Studies.
10	7	http://sun-volt.com This website is now defunct.

POLAR BEARS IN THE HOT TUB

10	8	https://en.wikipedia.org/wiki/Liddell_Power_Station Wikipedia article on Liddell Power Station
12	9	https://commons.wikimedia.org/w/index.php?curid=28787531 Wikimedia image under public domain.
12	10	https://www.youtube.com/watch?v=_oAap2XZdoc YouTube video of German Air Force's F-4 fly-by
16	11	https://www.jasonbox.net/ Images of black ice from research funded by The Geologic Survey of Denmark and Greenland
19	12	Data from book by Susan J Crockford ISBN: 1541123336
20	13	https://www.ncdc.noaa.gov/sotc/global/201806 NOAA provided graph of Temperature Anomalies - 1880 to 2008
20	14	https://oceanexplorer.noaa.gov/explorations/09ecs/media/ibcaoimage.html NOAA image. Circles locating islands and ridges added by Author.
24	15	Chart of CO2 level, Air Temperature, Arctic Ice Areas and Earthquake energy. Created by the Author with data from NOAA and other data.
25	16	The 40,000-MileVolcano By William J Broad Universität Bremen, Center for Marine Environmental Sciences
26	17	https://www.amap.no/documents/doc/surface-ocean-currents-in-the-arctic/566 Arctic Ocean Currents; Monitoring and Assessment
32	18	Monthly Weather Review, Nov 1922 By Nicholas Ifft. Arctic Ice Melt by 1922.

30	19	Author's photo of The GEORGE F. CRAM COMPANY'S IMPERIAL GLOBE. South Pole Area. Red dot location of volcanoes added by Author.
34	20	https://en.wikipedia.org/34/List_of_volcanoes_in_Antarctica Wikipedia article listing 36 Antarctic Volcanoes known before 2016
36	21	Table of Density of Exhaust Gas from Different Fuels Created by the Author.
38	22	https://www.aviationweather.ws/095_Thermal_Soaring.php From "full text of the classic FAA guide"
42	23	Created by the Author with data from ice core samples and Mauna Loa data.
47	24	https://www.ncdc.noaa.gov/sotc/global/201806 NOAA provided graph of Temperature Anomalies - 1880 to 2008
48	25	Created by the Author with data from Earthquake Watch, Charctic Interactive Sea Ice Graph and sources referenced earlier.
53	26	https://www.ncdc.noaa.gov/sotc/global/201806 NOAA provided graph of Temperature Anomalies - 1880 to 2008
60	27	https://forum.arctic-sea-ice.net/index.php/topic,1461.msg68601.html#msg68601 Image from open forum blog post by diablobanquisa on Arctic Sea Ice Forum
61	28	https://www.esa.int/spaceinimages/Images/2005/06/Earth's_wind_patterns Image available under ESA Standard License.
79	29	In 1922, the US Weather Bureau, now part of NOAA provided this article to the Boston Globe. Reported by George Nicholas Ifft.

POLAR BEARS IN THE HOT TUB

Arthur Krugler is a licensed chemical and mechanical engineer in five states. He is one of the leading geothermal engineers in the world.

Yogi Berra said you can see a lot by observing. That is one of my core beliefs and traits. This book is about observing the NOAA and other data. Observations lead to conclusions different than most published conclusions.

Management, Organizations & Education
Past director of the Geothermal Research Council.
Past president of the Ben Holt Company.
Member of AIChE for 50+ years.
BS Chemical Engineering University of Wisconsin.
Past director of the Gardina City, CA Chamber of Commerce

International
Worked on projects in New Zealand, Kenya, Indonesia, Peru, Canada, Domenica, and Tibet.

National
Licensed chemical and mechanical engineer in California, Nevada, Arizona, Hawaii, and Utah.

Industries
Geothermal, solar, gas and steam turbines, hydroelectric, oil, chemical and plastics industries.

Patents
Granted patents for graphite production, heat transfer and molding equipment.

He lives in California with his wife.

www.ingramcontent.com/pod-product-compliance
Lightning Source LLC
Chambersburg PA
CBHW041101180526
45172CB00001B/58